A, B, See... in 3D
A Workbook to Improve 3-D Visualization Skills

Dan G. Dimitriu, PhD, P.E.
San Antonio College

A, B, See.... in 3D: A Workbook to Improve 3-D Visualization Skills
Dan G. Dimitriu

ISBN-978-3-031-00939-6   paperback

ISBN-978-3-031-02067-4   ebook

First Edition

10 9 8 7 6 5 4 3 2 1

Contents

ABSTRACT

The workbook provides over 100 3D visualization exercises challenging the student to **create three dimensions from two.** It is a powerful and effective way to help engineering and architecture educators teach spatial visualization. Most of the 3-D visualization exercises currently being used by students in Design and Graphics classes present the objects in isometric views already in 3-D, asking the viewer to create multiple views, fold patterns, manipulate, reflect, or rotate them. The exercises presenting the objects in incomplete multiview projections asking the students to add missing lines use mostly real 3D objects that are more easily recognizable to help the student correlate 2D with 3D.

This workbook uses a different approach. Each view of the solid represents a letter of the alphabet. The letters are by definition 2D representations and when they are combined to create a 3D object, visualizing it becomes quite a challenge.

This workbook is intended for Engineering, Architecture, and Art students and faculty that want to increase their 3-D visualization skills.

Introduction

There is ample evidence that instruction in spatial visualization skills is effective in improving outcomes for engineering students. Research conducted since the early 1990's has proven that spatial visualization practice and training leads to better grades in engineering graphics and engineering coursework, and in the retention of underrepresented groups in the field.

In 1993 Dr. Sheryl Sorby (1998, 2006, 2009) of Michigan Technological University began work under and NSF grant to figure out why first-year women engineering students were more than three times as likely to fail the Purdue Spatial Visualization Test: Rotations (PSVT:R) as men, and what could be done about it. Sorby's analysis of the results of the test and a background questionnaire she administered to test-takers showed that previous experience in design-related courses such as drafting, mechanical drawing, and art, as well as play as children with construction toys such as Legos, Lincoln Logs, and Erector Sets, predicted good performance on the PSVT:R. She and her colleagues then developed a three-credit hour spatial-visualization course and administered it to students who had failed the PSVT:R. The course covered topics such as cross sections of solids, sketching multiview drawings of simple objects, and paper folding to illustrate 2-D to 3-D transformations. In the lab, students used solid-modeling computer-aided design (CAD) Software. Student's tests scores on the PSVT:R improved from an average of 52% to an average of 82%. Work by Hsi, et al (1997) supported the effect of spatial strategies instruction on erasing gender differences and improving grades for engineering students.

Additional spatial visualization training was also discovered to positively affect retention in engineering for women. Sorby found that among the women who initially failed the PSVT:R and took the spatial-visualization course between 1993 and 1998, 77% were retained in Engineering Design, compared to 48% of the women who didn't take the course (Female n=251). In additional studies Sorby also found that middle school girls who took a spatial-visualization course took more advanced-level math and science courses in high school than did girls who did not take the course, and that the materials were shown to be effective in improving spatial skills for undergraduate students outside of engineering, and for students in high school.

After offering the three-hour spatial visualization course for six years (yielding improvements of 20 to 32 percentage points on the PSVT:R), Sorby condensed the course to a one-hour course and tested it between 2000 and 2002, seeing average improvement on the PSVT:R of 26%. In 2003 Sorby, Beverly Baartmans and Anne Wysocki, published a multimedia software-workbook package which contained content similar to the course *Introduction to 3D Spatial Visualization*, now used for engineering graphics education throughout the nation.

At Penn State Erie, Dr. Kathy Holliday-Darr and Dr. Dawn Blasko (2009) conducted a one-credit-hour intervention with mechanical engineering technology and plastic engineering technology students who performed below criteria on the PSVT, the Mental Rotation Test (MRT, and paper-folding and water-level tasks. They used the Sorby and Wysocki multimedia package and found significant improvement compared to an untreated control group. The improvement was correlated with grades in other courses and scores on spatial tasks correlated with overall GPA and key courses taken in the following semester and year.

At Virginia State University, a Historically Black College or University (HBCU), retention of minorities in STEM-related majors tended to be lower than their non-minority peers, and students enrolled in introductory engineering graphics courses had significantly lower-than-average test scores on the PSVT. Dr. Nancy Study piloted changes to engineering graphics courses, including the use of sketching, blocks and multimedia, that resulted in improvement of students' visualization abilities. Significantly higher GPAs were earned by students taking the enhanced pilot engineering graphics course, compared to a control group that did not take the enhanced course, and a higher percentage of students in the test group were retained both in an engineering or technology major and at the university even if they did change their major.

Uttal, Meadow and Newcombe, (2010) conducted a meta-analysis of 200 studies on improvement of spatial skills and found that the average effect size of improvement for students who receive extensive practice on spatially-relevant skills, such as mentally rotating figures or disembodying, was .53 (equivalent to an intervention improving SAT scores by more than 50 points or IQ scores by more than 7.5 points). They also found that the effects of training endured over time, after practice interventions were completed.

Although the materials currently being used nationally are now assisting the new generations of engineering students to succeed, they do not challenge the student to **create three dimensions from two.** On today's market there are some valuable tools with which engineering and architecture educators teach spatial visualization. Most of the 3-D visualization exercises currently being used for students in Design and Graphics classes present objects in isometric views already in 3-D, asking the viewer to create multiple views, fold patterns, manipulate, reflect, or rotate them. The materials presented in this workbook take a universally accepted 2-D flat pattern (a letter) and ask the viewer to mold it as part of a 3-D solid, in combination with two other flat-pattern letters from adjacent views.

This workbook is intended for Engineering, Architecture, and Art students and faculty that want to increase their 3-D visualization skills.

Methodology

The exercises use alphabet letters represented in standard multiview projections (front, top, right, left, or bottom side views). The 3-D object made up of the three letters, one in each view, has to be mentally assembled in 3-D, with no assistance from an isometric representation of the solution figure. The problems ask the solver to break out of the 2-D image of the letter and visualize the third dimension, the depth, or "Z axis", and combine with the other two letters from the other views of the 3-D object. The exercises are presented with increasing degrees of difficulty to help students improve their 3-D visualization skills. No other universally-recognized flat patterns are currently being used to enhance students' ability to spatially visualize 3-D objects.

"A, B, See... in 3D" presents over 100 three-letter combinations many of them with multiple solutions and a brief instructional text on how to solve these exercises. The problems will build on the body of knowledge already developed in early stages of graphics core concepts such as:

- Alphabet of lines (visible, hidden, and center lines)
- Multiview Orthographic Projections
- Surface and Edge Classifications (Normal, Inclined, Oblique, and Curved)

The graphical problems are presented in order of increasing difficulty and they are designed to gradually break students out of their 2-D preconceptions about 3-D space. The student must learn how to represent a variety of surfaces normal, inclined, oblique and cylindrical in multiple positions, visible and invisible, in edge view, true size and shape, or foreshortened in order to complete these assignments. No other universally-recognized flat patterns are currently being used to enhance students' ability to spatially visualize 3-D objects.

Figure 1

Solution: 9 cubes

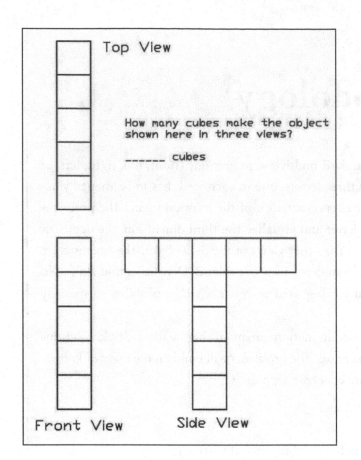

Top View

How many cubes make the object shown here in three views?

_____ cubes

Front View   Side View

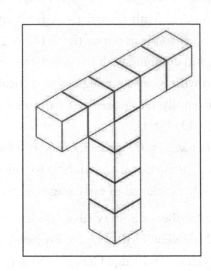

Figure 2

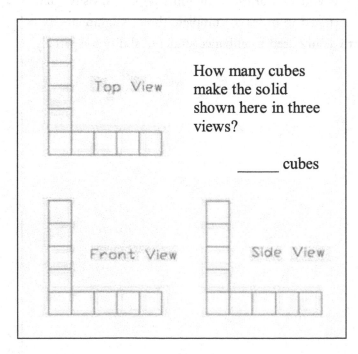

Top View

How many cubes make the solid shown here in three views?

_____ cubes

Front View   Side View

Solution: 13 cubes

Figure 3

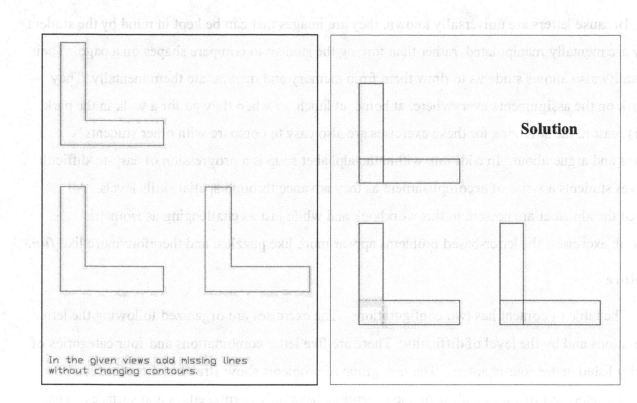

In the given views add missing lines
without changing contours.

Figure 4   **Figure 5**

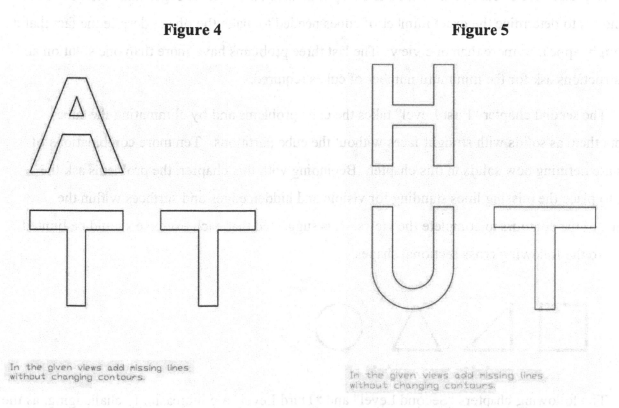

In the given views add missing lines
without changing contours.

In the given views add missing lines
without changing contours.

See solutions for Figures 4 and 5 on page 9.

7

Because letters are universally known, they are images that can be kept in mind by the student as they are mentally manipulated, rather than forcing the student to compare shapes on a page. Their universality also allows students to draw them from memory and manipulate them mentally. They can work on the assignments everywhere, at home, at lunch, or when they go for a walk in the park. For this reason, the solutions for these exercises are also easy to compare with other students' solutions and argue about. In addition, within the alphabet soup is a progression of easy-to-difficult that gives students a sense of accomplishment as they advance through spatial skills levels. All letters of the alphabet are present in this workbook and while just as challenging as isometric workbook exercises, the letter-based problems appear more like puzzles, and therefore more like *fun*.

Procedure

The table of content has two configurations. The exercises are organized following the letter combinations and by the level of difficulty. There are five letter combinations and four categories of difficulty listed under four chapters. The first group of problems show straight letters that can be made completely out of cubes and, with one exception, have only normal edges and surfaces. The challenge is to determine the exact number of cubes needed to make the object despite the fact that a cube might appear in more than one view. The last three problems have more than one solution and the instructions ask for the minimum number of cubes required.

The second chapter "First Level" takes the cube problems and by eliminating the cubes presents them as solids with straight faces without the cube partitions. Ten more combinations of letters are defining new solids in this chapter. Beginning with this chapter, the problems ask the solver to place the missing lines standing for visible and hidden edges and surfaces within the confines of the contours to complete the views. It is suggested that each exercise should be limited initially to the following cross sectional shapes:

The following chapters "Second Level" and "Third Level" are increasingly challenging, as the solvers have to visualize letter parts that are not where they are appear to be from the 2D image. Almost all the problems have multiple solutions as the letter parts can be visualized from rectangular prisms, to triangular ones, or even cylindrical shapes as indicated above. At these advanced stages students can experiment with other shapes as well such as a hexagon or a rhombus. The challenge is to have the correct representation with visible, hidden, and centerlines in each view. The advantage of all these exercises is that the solution can be easily verified for correctness by building a 3D model in any 3D modeling package. The standard front, top, and side view projections of the model should reveal all the necessary lines for verification.

Many times the students start to look for new combination of letters which is a challenge in itself because not all combinations of three letters can form a solid. That is another way to improve the visualization abilities.

Imagination is the only limit!

Solutions for Figures 4 and 5:

9

List of Problems by Difficulty

Alphabetical listing by Letter Combinations

Letters Made Out of Cubes

EFT	C 09
HHH	C 10
HLE	C 08
HLL	C 05
HTT	C 11
IIH	C 03
IIL	C 02
IIT	C 01
LLL	C 04
LUF	C 12
TTT	C 06
UHL	C 07

Three Identical Letters

DDD	F 21
EEE	S 22
FFF	S 23
HHH	F 11
LLL	F 06
OOO	F 19
TTT	F 10
XXX	T 21

Two Identical + One Different

EEN	S 13
IIA	F 04
IIH	F 03
IIL	F 02
IIM	F 05
IIT	F 01
LLA	F 09
LLD	F 08
LLQ	S 03
TTH	F 15

One Different + Two Identical

ATT	F 14
HLL	F 07
MOO	T 10
OHH	F 16
XAA	T 15
XTT	F 13
XVV	T 16
ZEE	T 07
ZHH	F 17
ZOO	T 09

All Three Different Letters

AUL	F 26
DEL	F 20
EDU	S 12
EFD	S 14
EFT	F 12
ELM	S 15
FEZ	T 01
FJT	T 02
GOP	T 03
HEB	T 18
HLE	F 18
HMT	S 01
HUT	S 16
HUV 1	S 17a
HUV 2	S 17b
HZT	S 24
KLE	S 05
LCX	T 04
LED	S 02
LOT	S 04
LUF	F 24
MHE	T 05
MHF	T 06
MLU	S 18
MOE	S 19
MUD	S 20
MUE	S 21
MUG	S 22
MUL	S 23
MUZ	S 30
NHB	T 17
NUE	S 26
NUL	S 25
OLE	S 27
OUA	T 11
PET	S 06
PFT	S 08
PLE	S 09
POT	S 10
RET	S 07
SET	S 11
TAP	T 19
UAL	S 28
UHL	F 25
VLT	S 29
WMX	T 20
WTF	T 12
YTF	T 13
ZEF	T 08
ZXN	T 14

Chapter 1 Problems - Combination of Cubes (C)
Level of difficulty: **Easy**

How many cubes make the object shown here in three views?

Top View

______ **cubes**

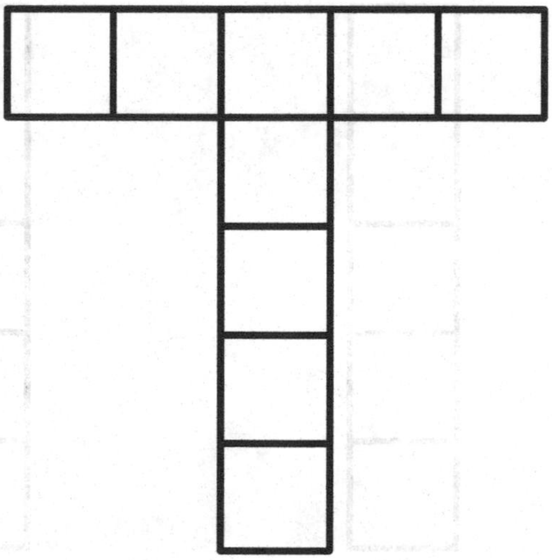

Front View

Side View

C 01 | **Name:**

How many cubes make the object shown here in three views?

Top View

_______ cubes

Front View

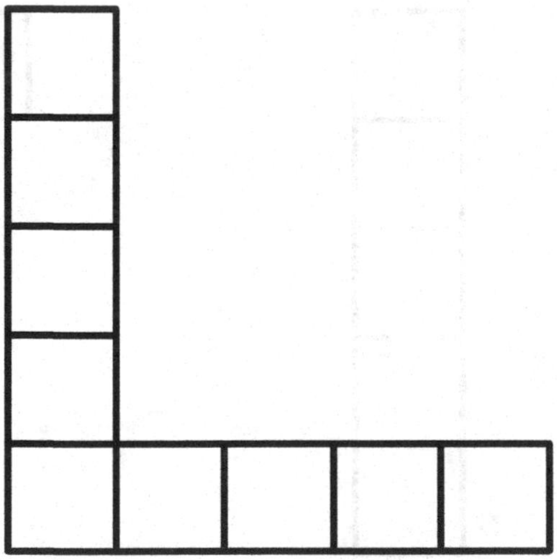

Side View

C 02	Name:

How many cubes make the object shown here in three views?

Top View

_____ cubes

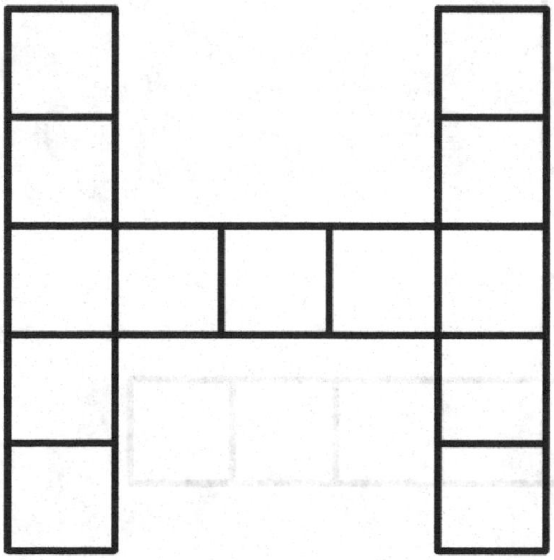

Front View

Side View

Top View

How many cubes make the object shown here in three views?

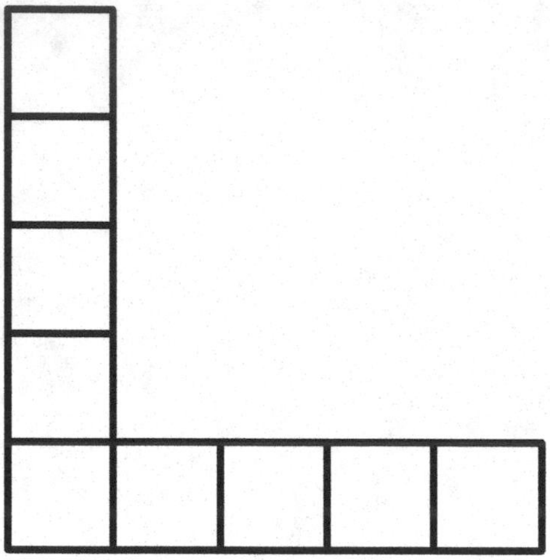

_____ cubes

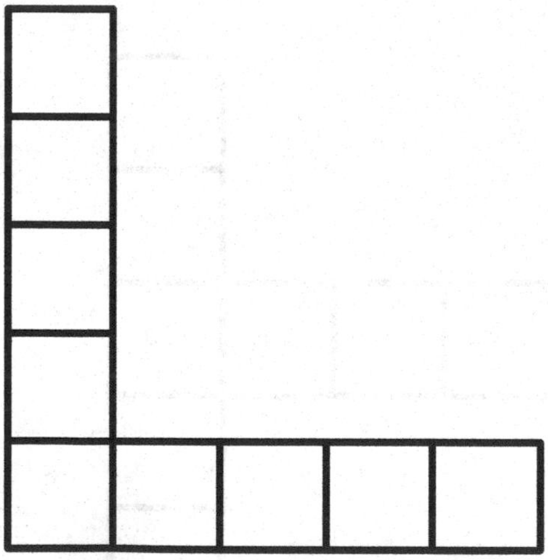

Front View

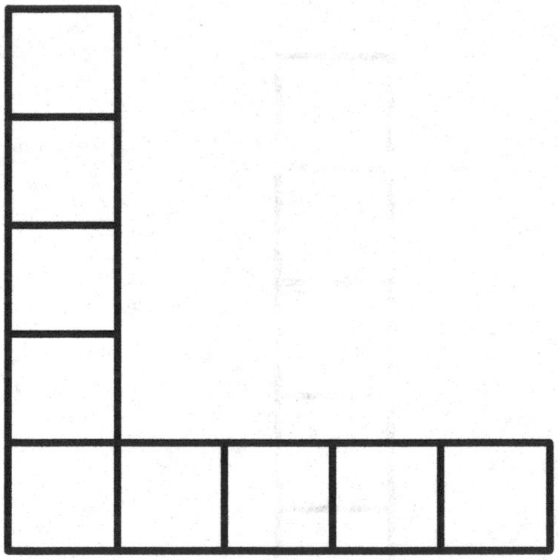

Side View

Top View

_____ cubes

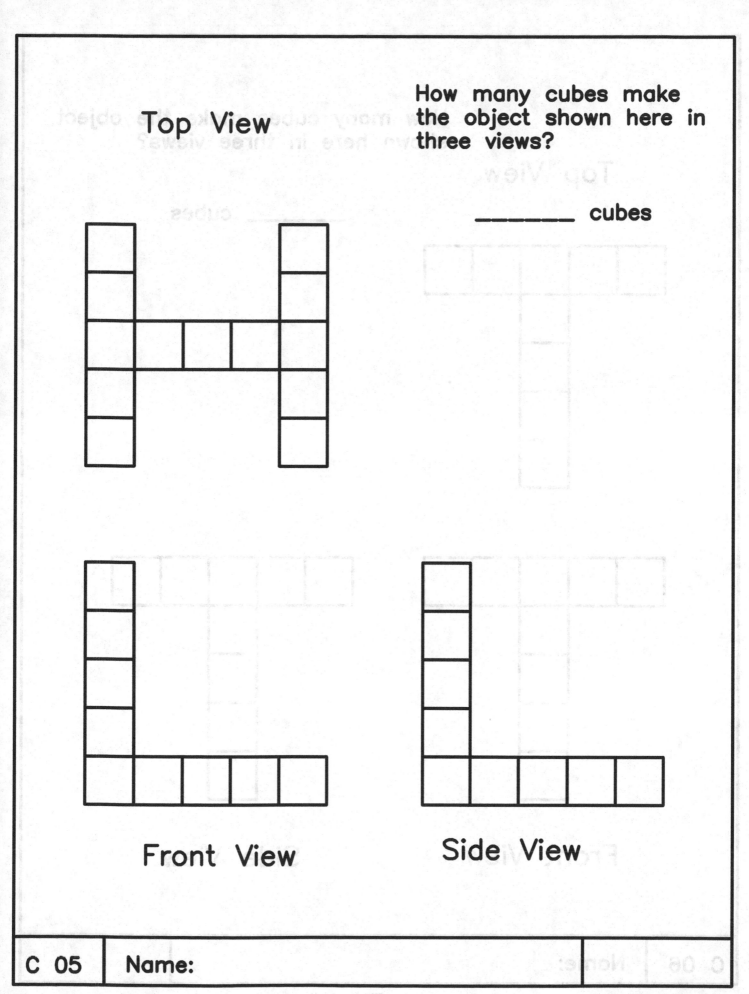

Front View   Side View

C 05 | Name:

17

How many cubes make the object shown here in three views?

Top View

______ cubes

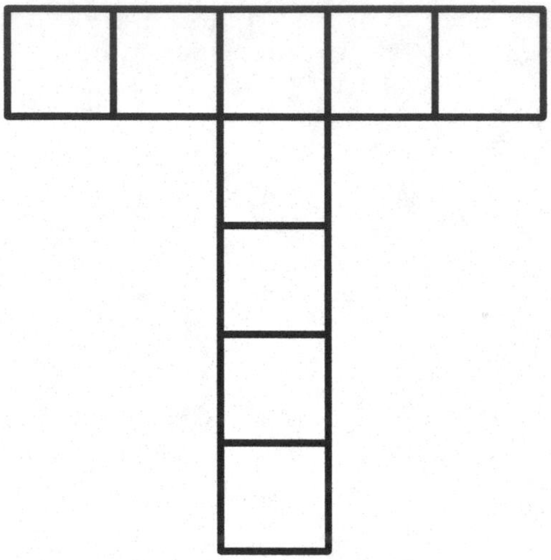

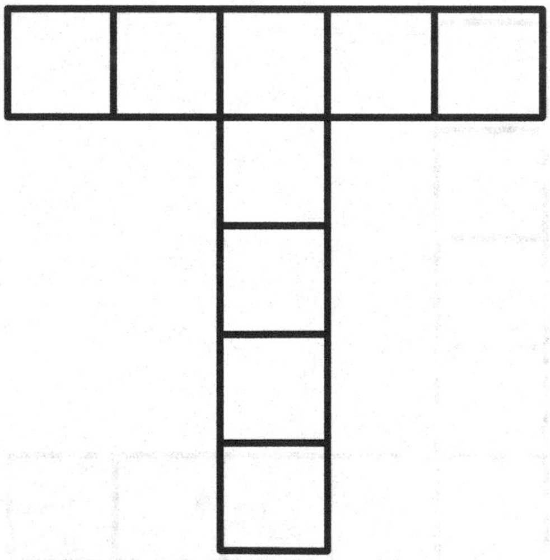

Front View

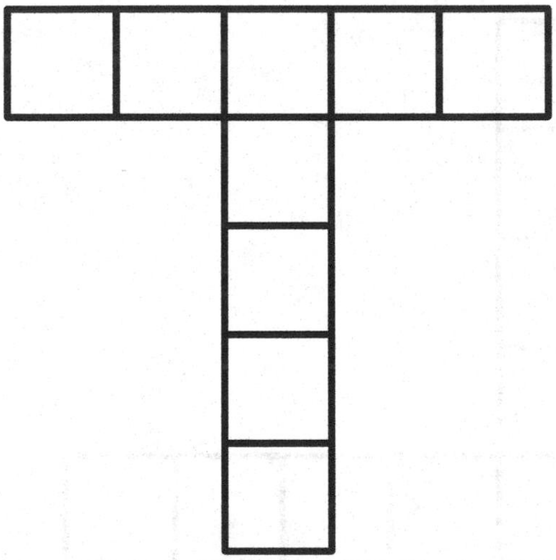

Side View

C 06	Name:

Top View

How many cubes are needed
to make the object shown here
in three views?

_____ cubes

Front View

Side View

Top View

How many cubes make the object shown here in three views?

_______ cubes

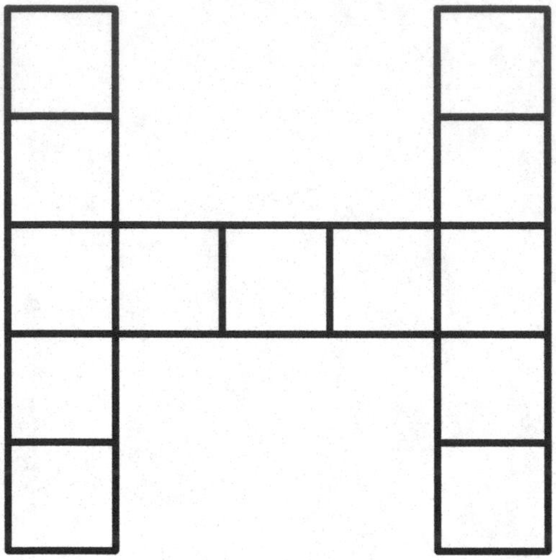

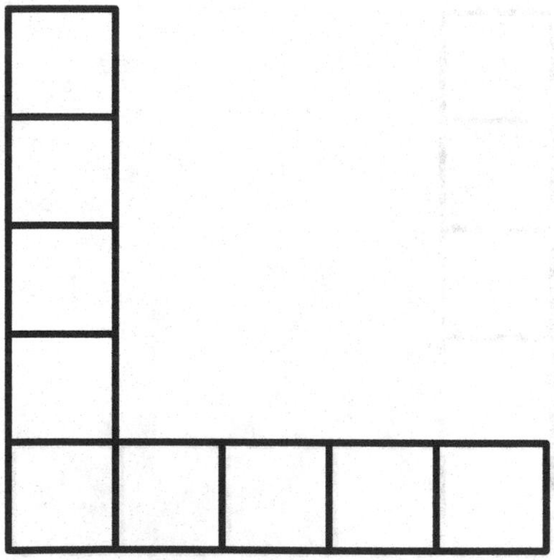

Front View

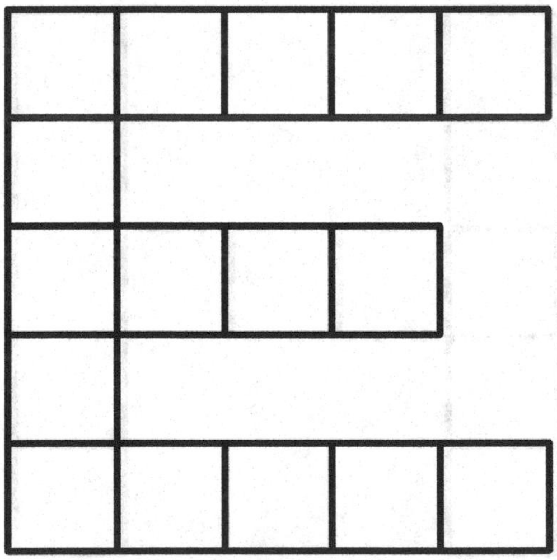

Side View

C 08 Name:

How many cubes make the object shown here in three views?

Top View

_____ cubes

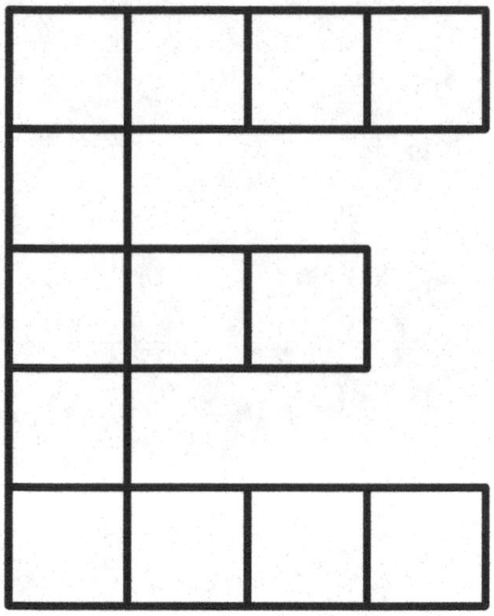

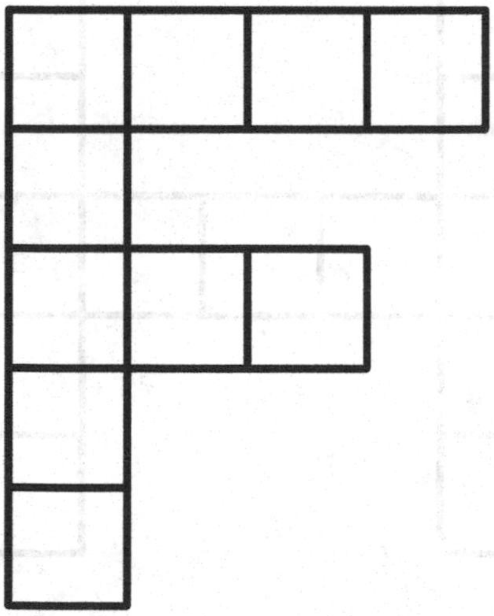

Front View

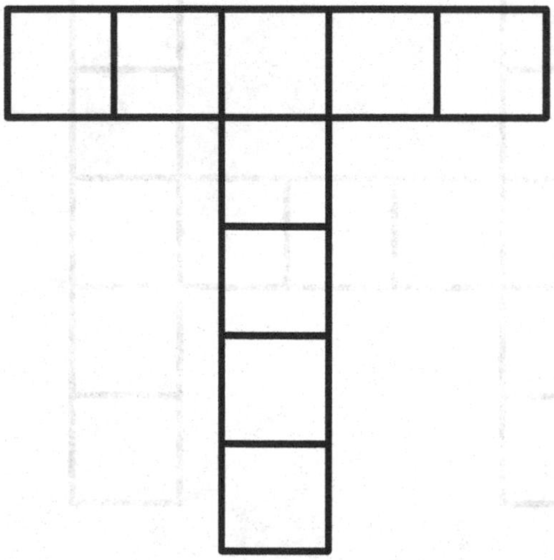

Side View

Top View

How many cubes make the object shown here in three views?

_____ cubes

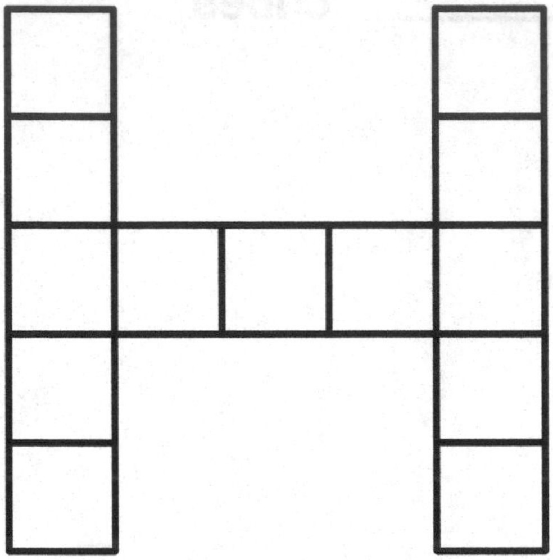

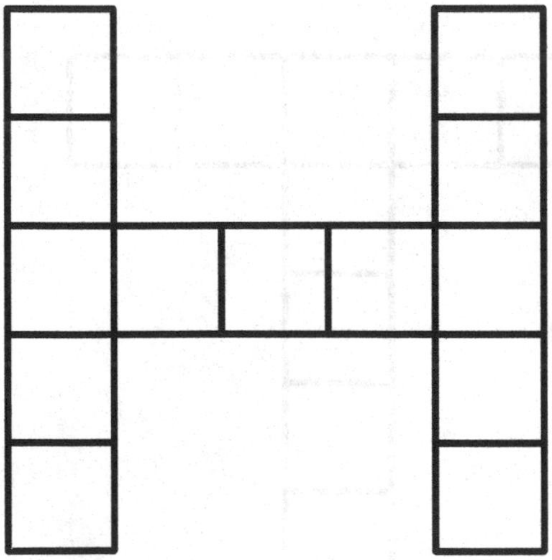

Front View

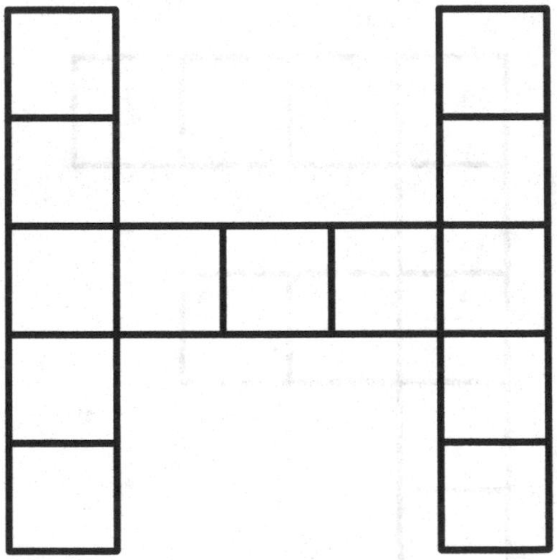

Side View

Top View

_____ cubes

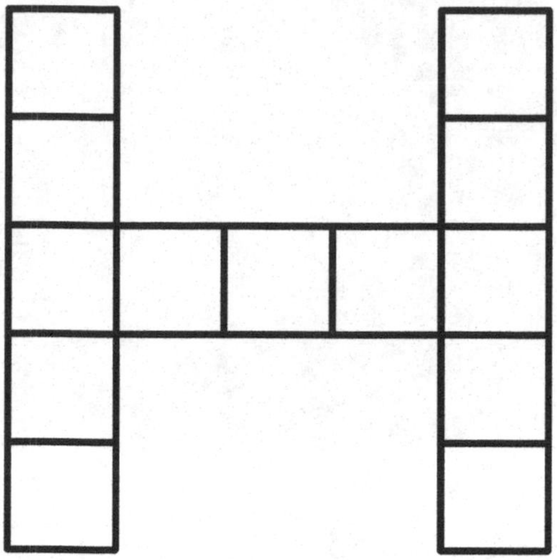

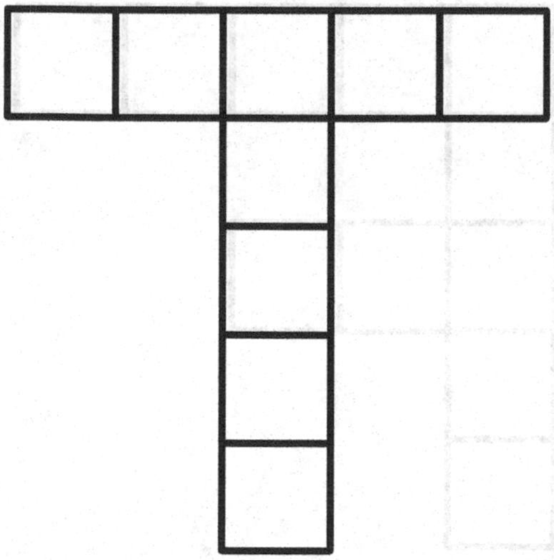

Front View

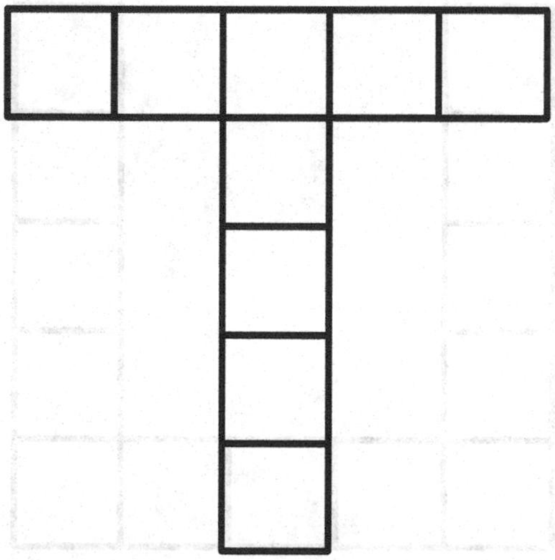

Side View

C 11 | Name:

23

How many cubes are needed to make the object shown here in three standard views?

______ cubes

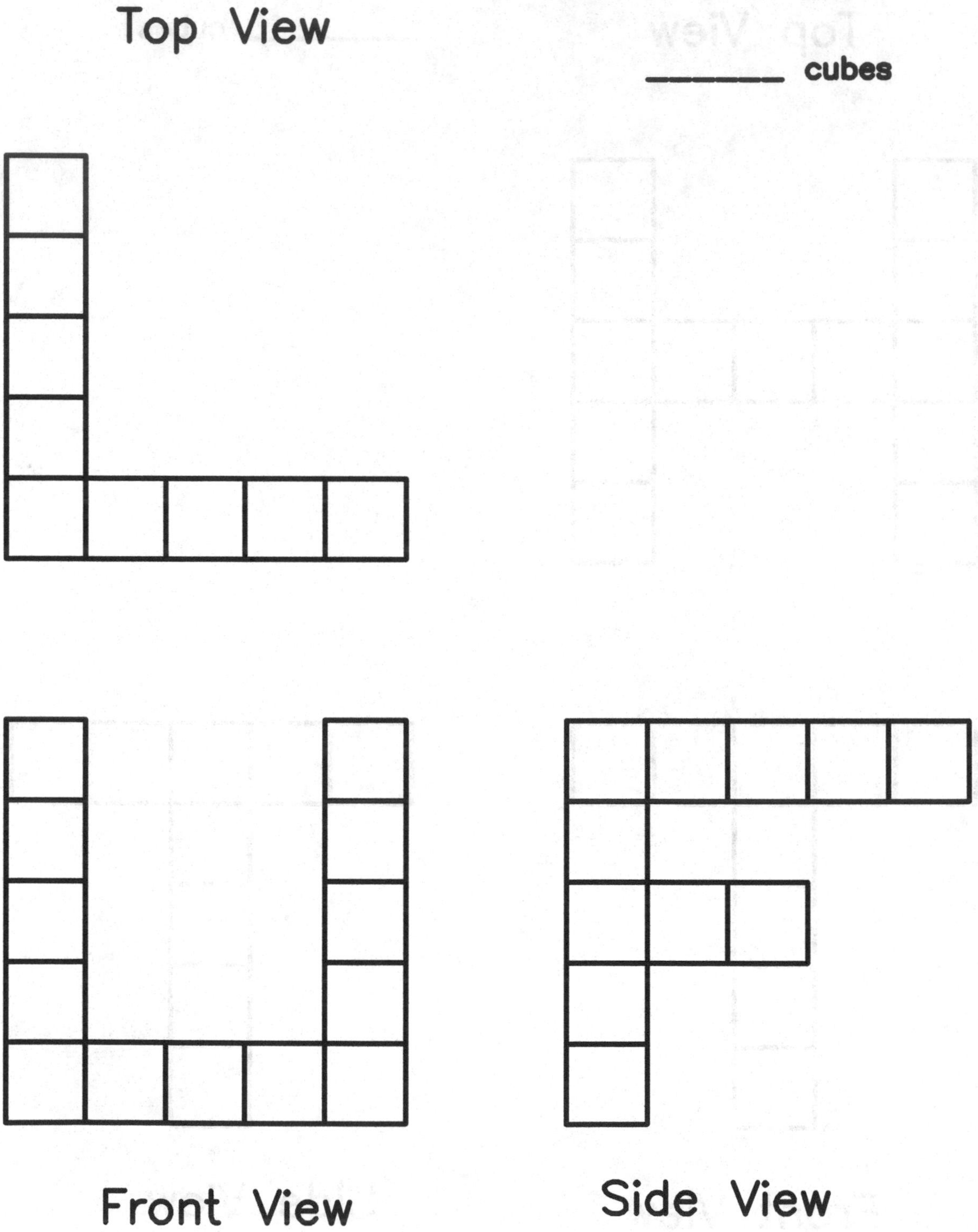

Top View
Front View
Side View

C 12 | Name:

Chapter 2 - First Level (F)

Level of difficulty: **Low**

Acceptable cross—sectional shapes:

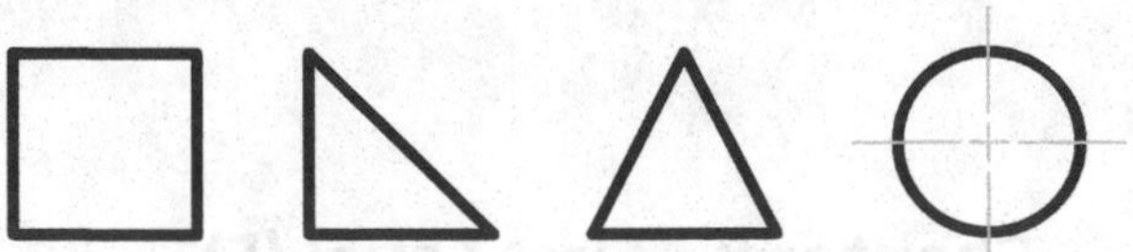

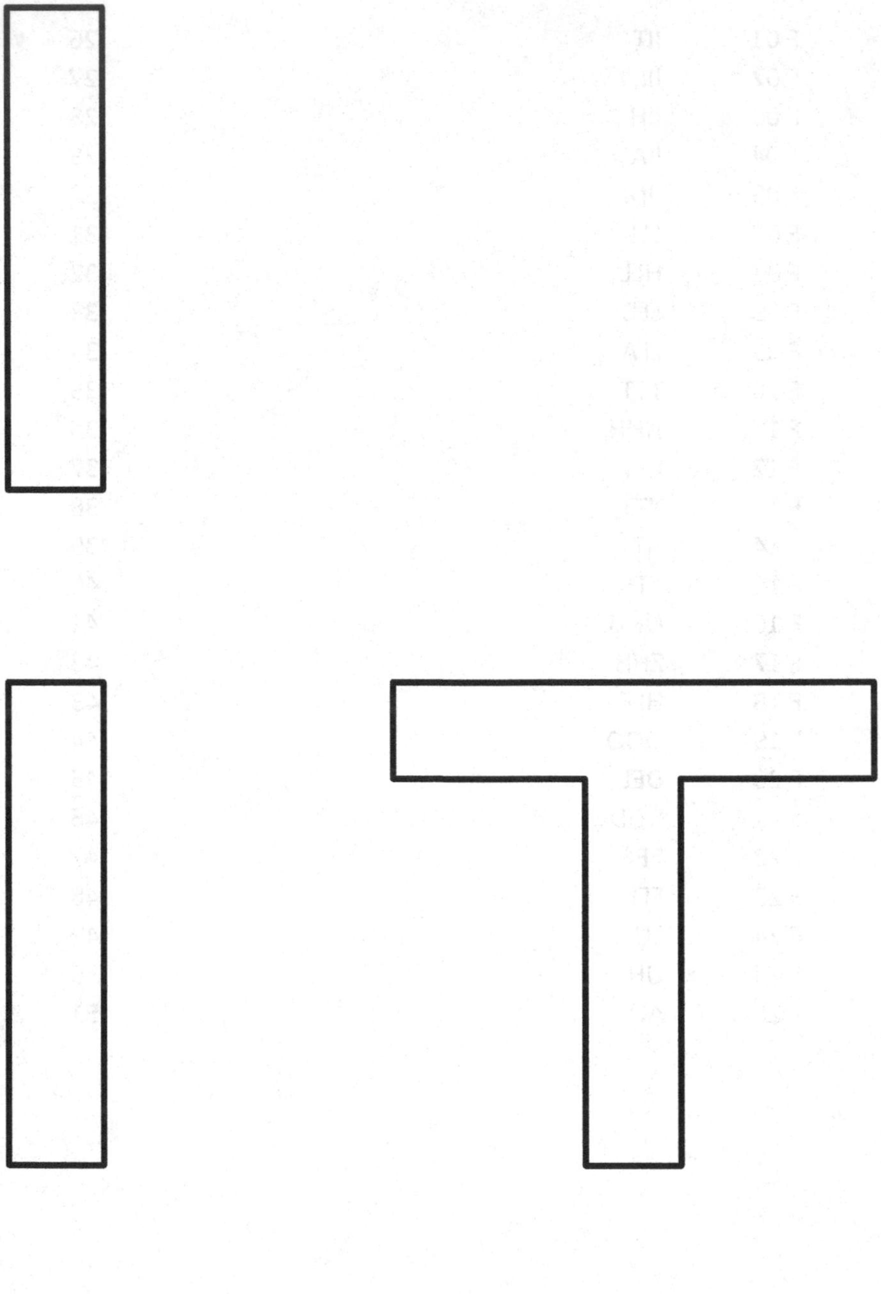

In the given standard orthographic views of the 3D solid object add the missing lines without changing contours.

F 01	**Name:**	

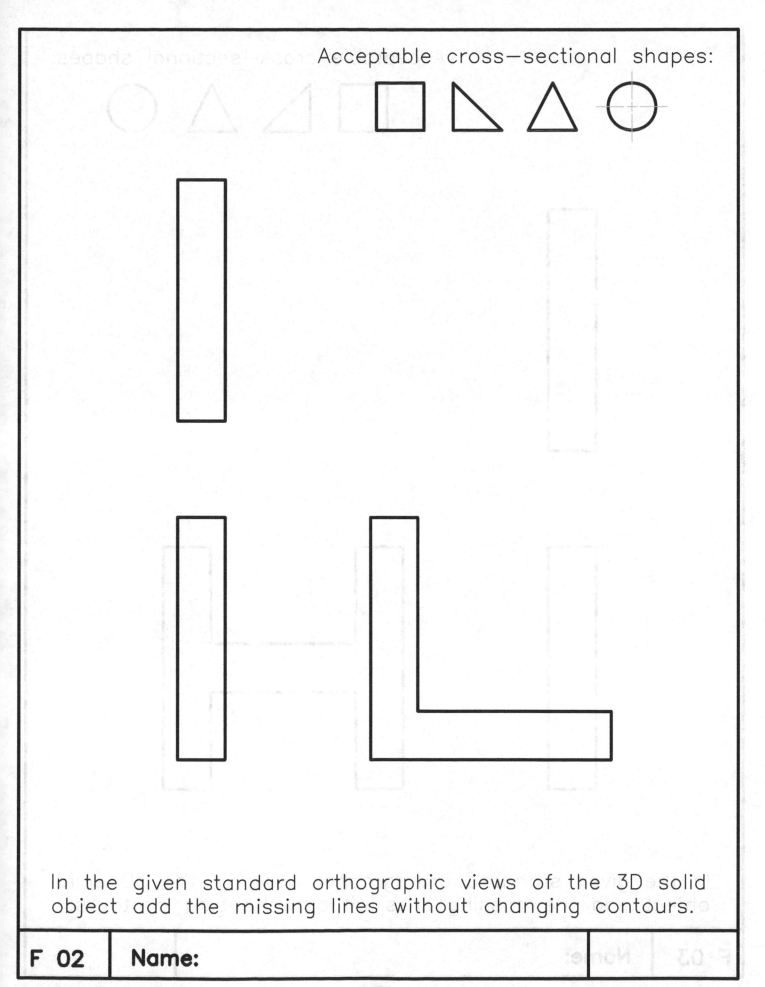

Acceptable cross—sectional shapes:

In the given standard orthographic views of the 3D solid object add the missing lines without changing contours.

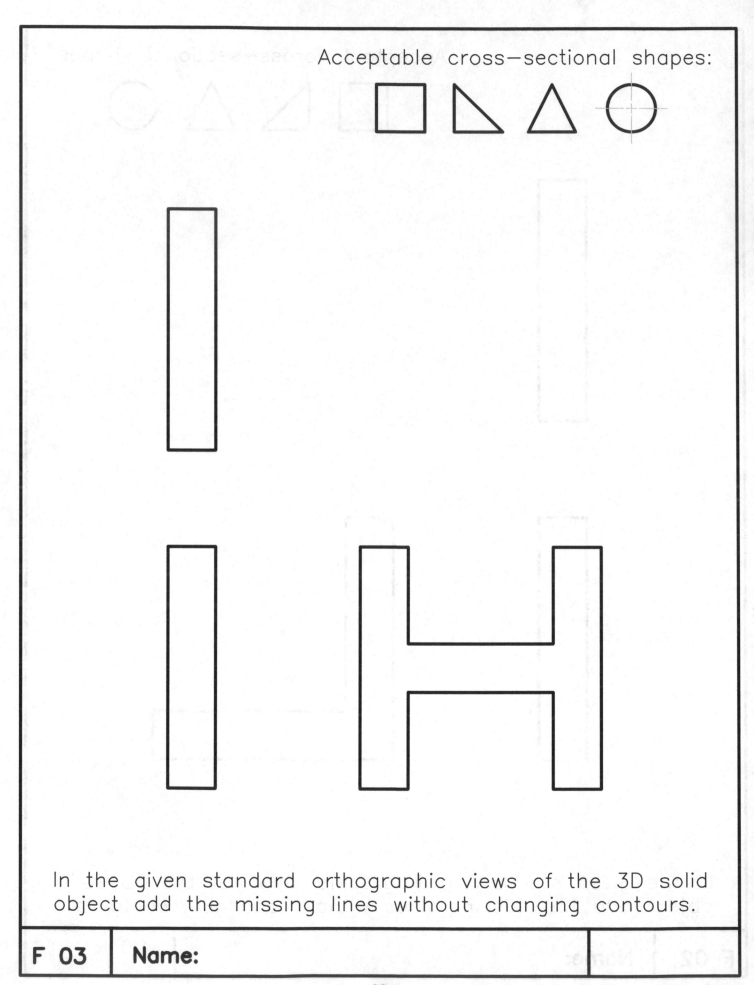

In the given standard orthographic views of the 3D solid object add the missing lines without changing contours.

F 03 | Name:

In the given standard orthographic views of the 3D solid object add the missing lines without changing contours.

F 04   Name:

Acceptable cross–sectional shapes:

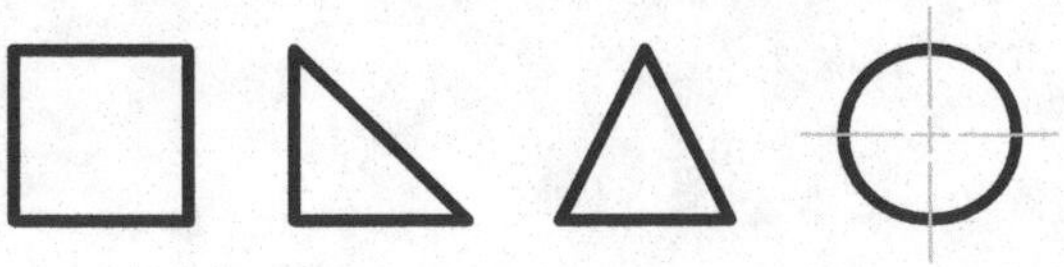

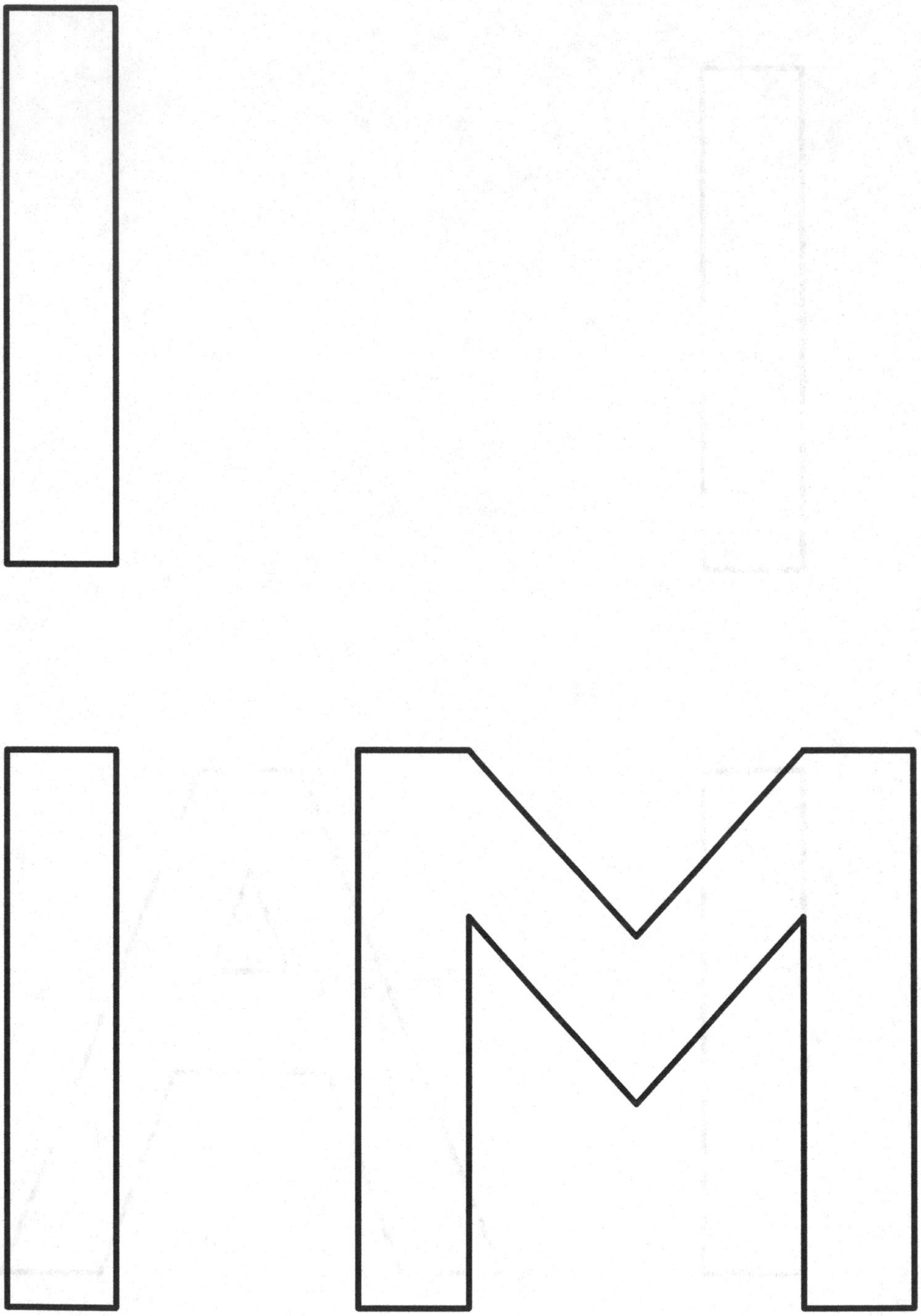

In the given standard orthographic views of the 3D solid object add the missing lines without changing contours.

F 05	Name:	

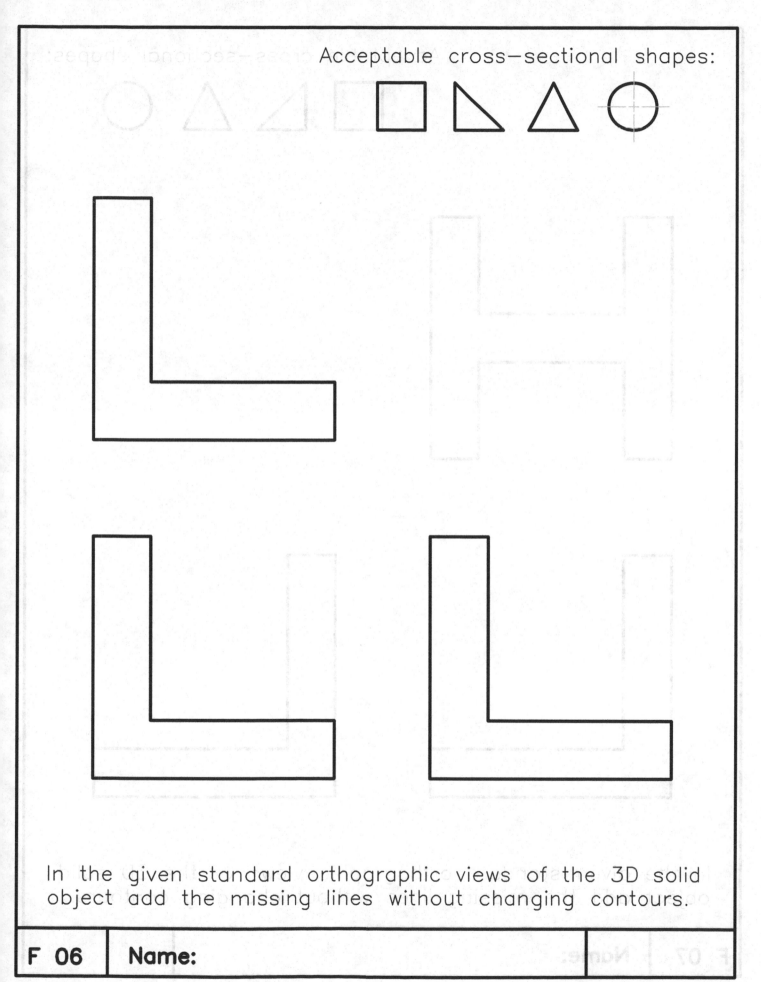

In the given standard orthographic views of the 3D solid object add the missing lines without changing contours.

F 06 | Name:

Acceptable cross-sectional shapes:

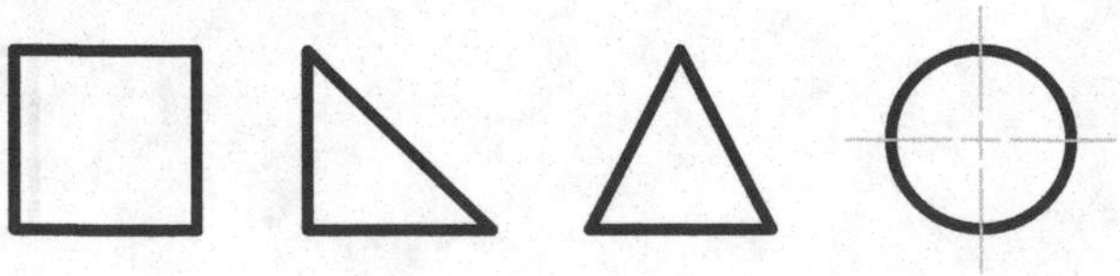

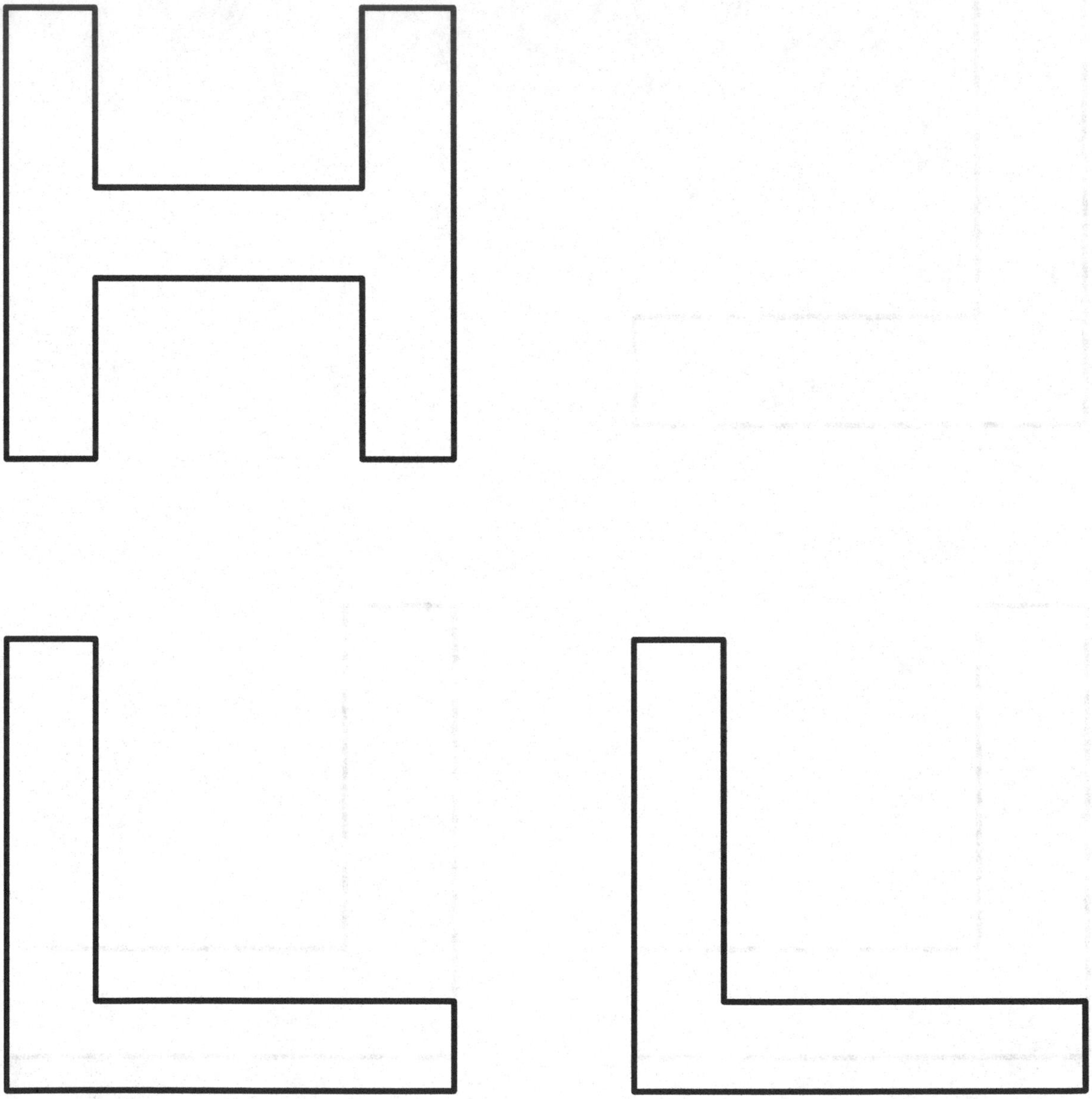

In the given standard orthographic views of the 3D solid object add the missing lines without changing contours.

F 07	**Name:**	

Acceptable cross—sectional shapes:

In the given standard orthographic views of the 3D solid object add the missing lines without changing contours.

In the given standard orthographic views of the 3D solid object add the missing lines without changing contours.

F 09   Name:

Acceptable cross—sectional shapes:

In the given standard orthographic views of the 3D solid object add the missing lines without changing contours.

F 10   Name:

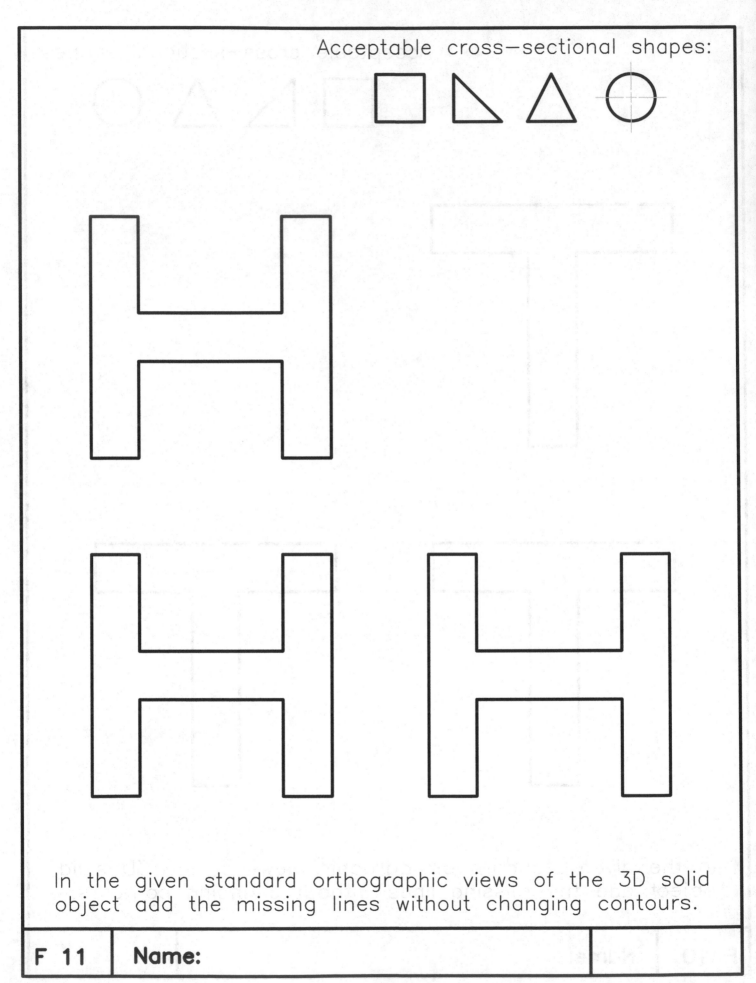

Acceptable cross—sectional shapes:

In the given standard orthographic views of the 3D solid object add the missing lines without changing contours.

F 11   Name:

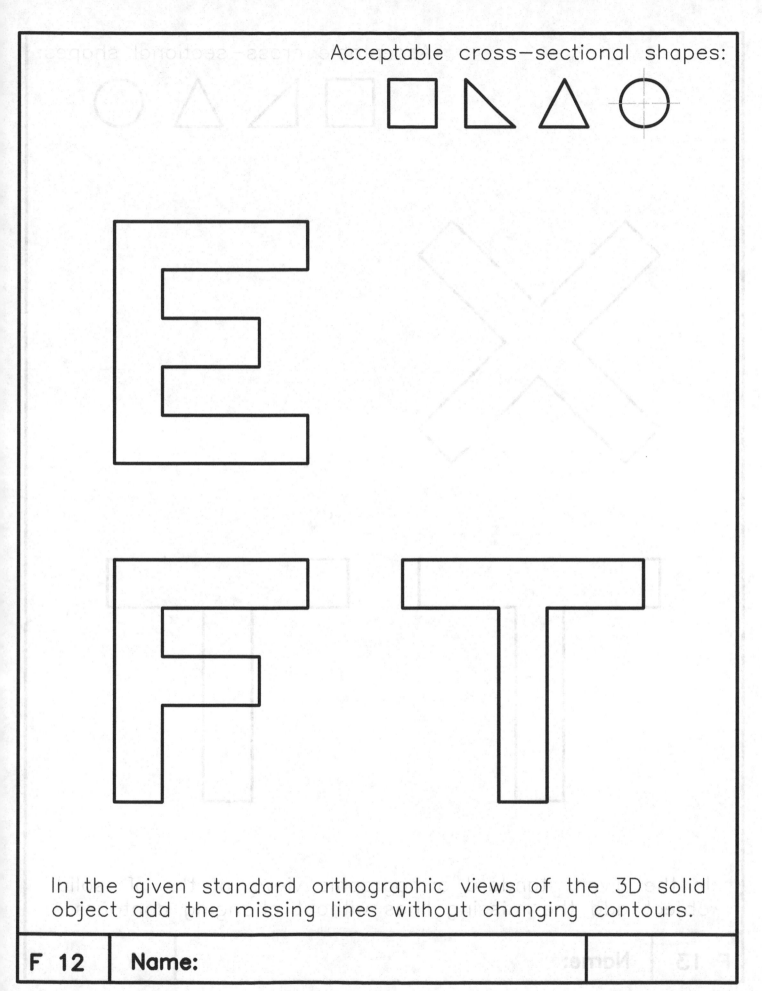

In the given standard orthographic views of the 3D solid object add the missing lines without changing contours.

| F 12 | Name: |

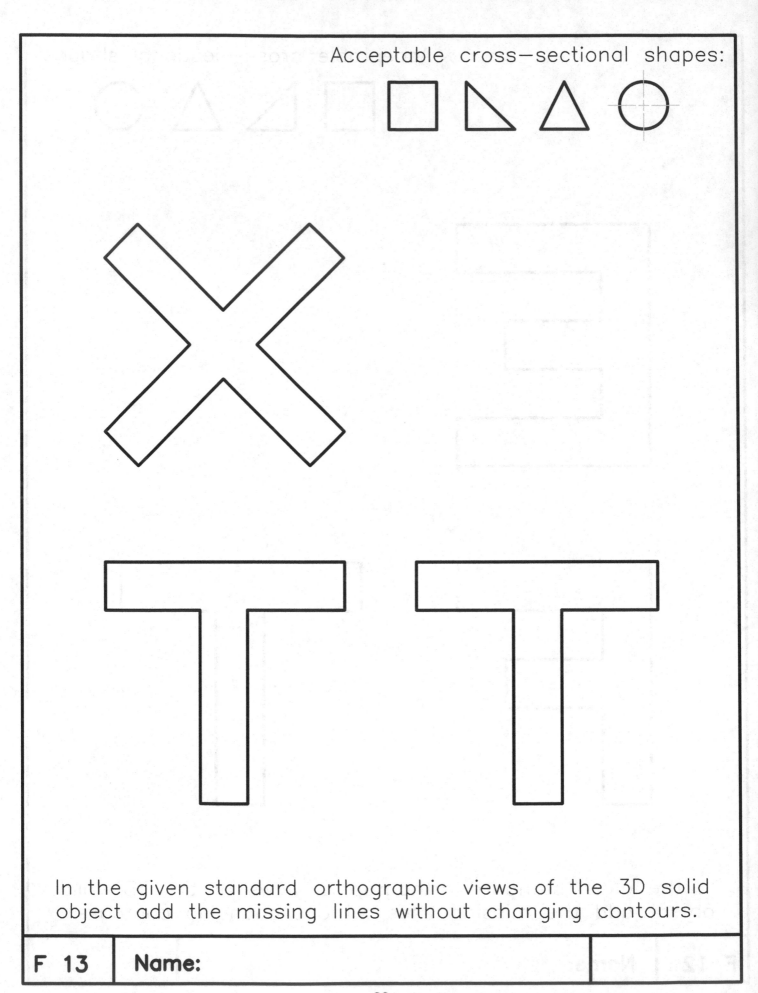

Acceptable cross—sectional shapes:

In the given standard orthographic views of the 3D solid object add the missing lines without changing contours.

F 13 | Name:

Acceptable cross-sectional shapes:

In the given standard orthographic views of the 3D solid object add the missing lines without changing contours.

Acceptable cross—sectional shapes:

In the given standard orthographic views of the 3D solid object add the missing lines without changing contours.

Acceptable cross-sectional shapes:

In the given standard orthographic views of the 3D solid object add the missing lines without changing contours.

F 16   Name:

41

In the given standard orthographic views of the 3D solid object add the missing lines without changing contours.

F 17   Name:

Acceptable cross-sectional shapes:

In the given standard orthographic views of the 3D solid object add the missing lines without changing contours.

F 18   Name:

43

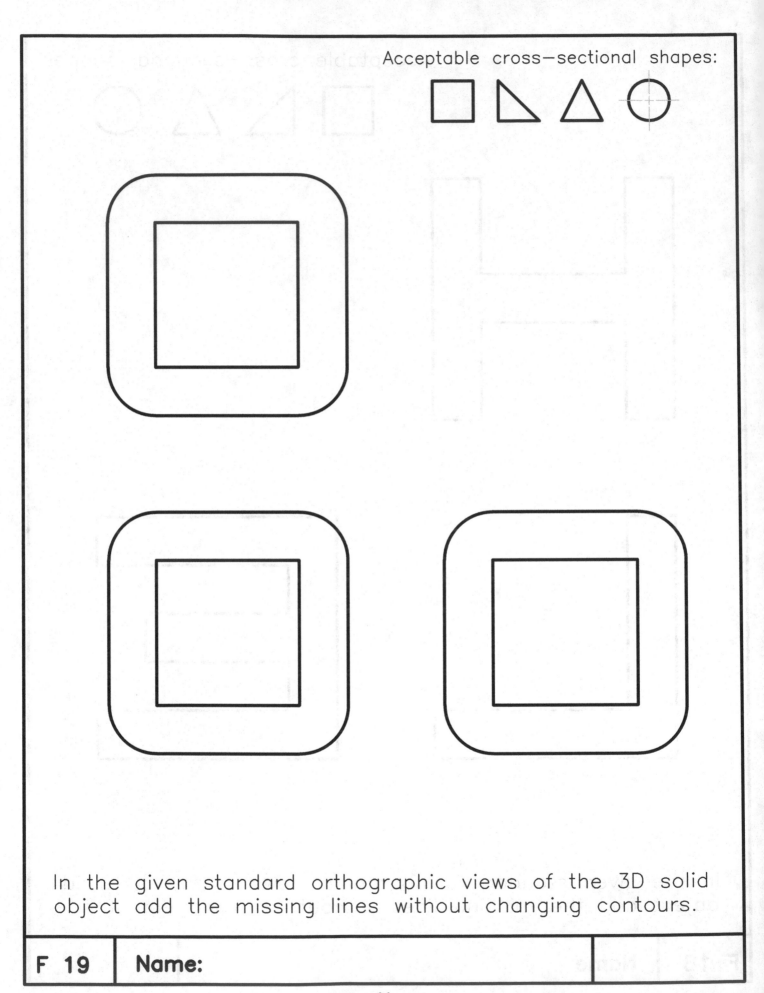

In the given standard orthographic views of the 3D solid object add the missing lines without changing contours.

F 19 | Name:

In the given standard orthographic views of the 3D solid object add the missing lines without changing contours.

F 20   Name:

In the given standard orthographic views of the 3D solid object add the missing lines without changing contours.

F 21 Name:

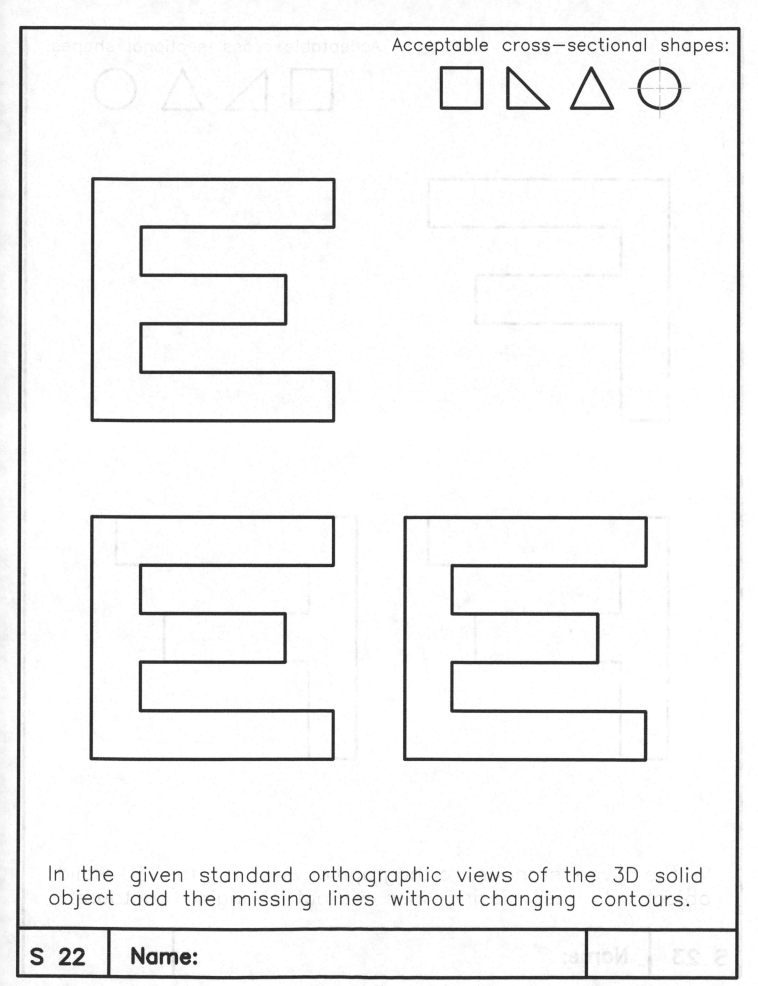

Acceptable cross—sectional shapes:

In the given standard orthographic views of the 3D solid object add the missing lines without changing contours.

S 22   Name:

Acceptable cross–sectional shapes:

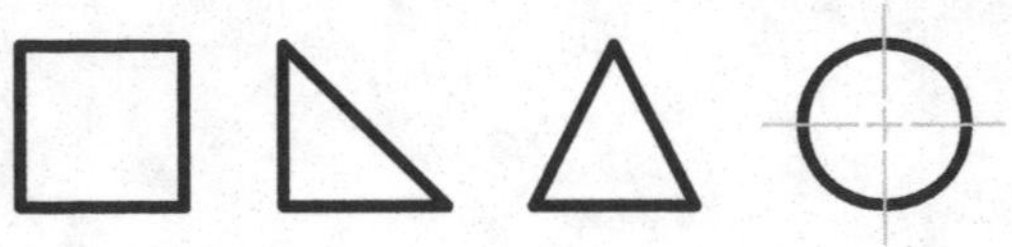

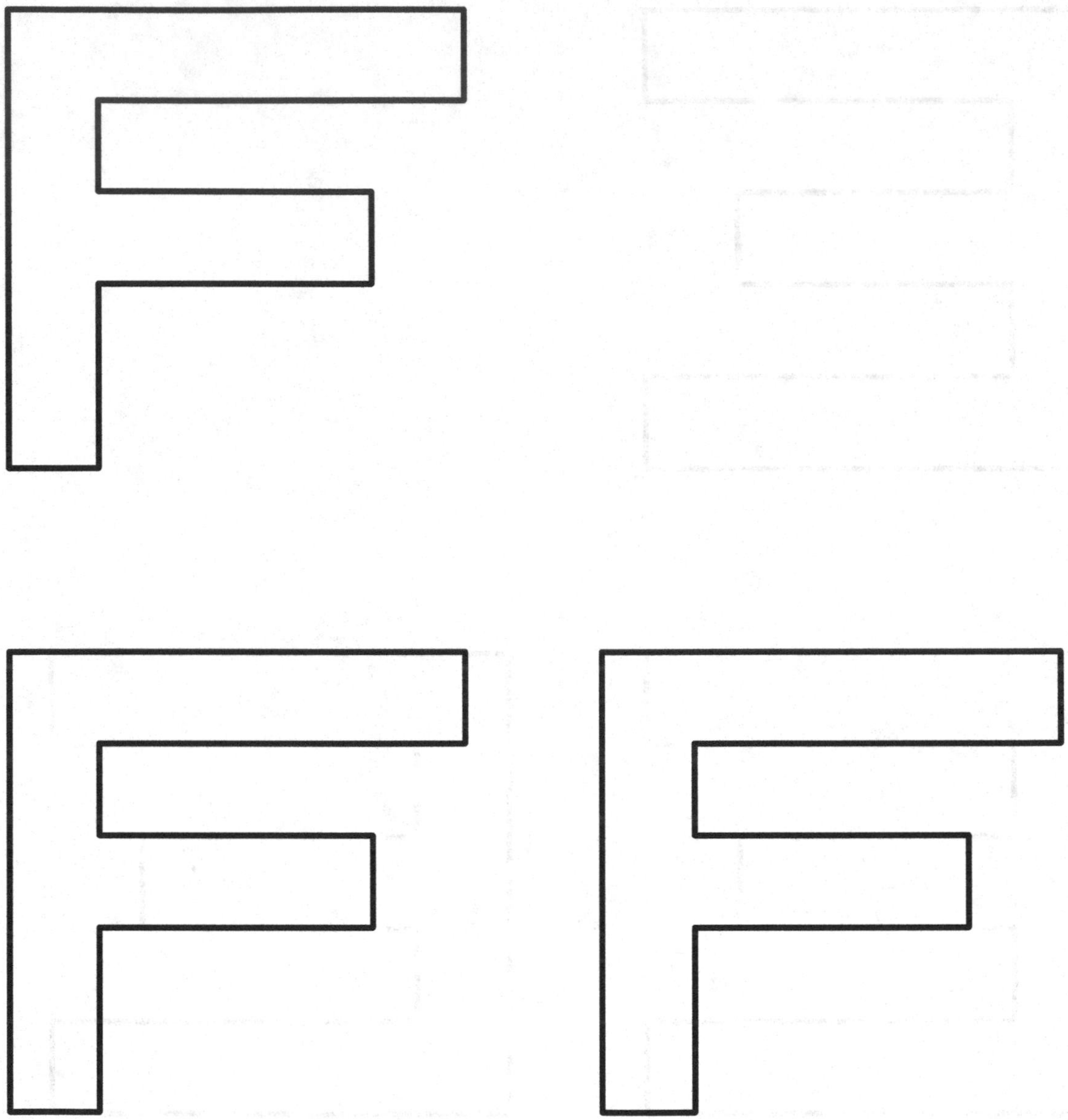

In the given standard orthographic views of the 3D solid object add the missing lines without changing contours.

S 23	**Name:**

In the given standard orthographic views of the 3D solid object add the missing lines without changing contours.

F 24   Name:

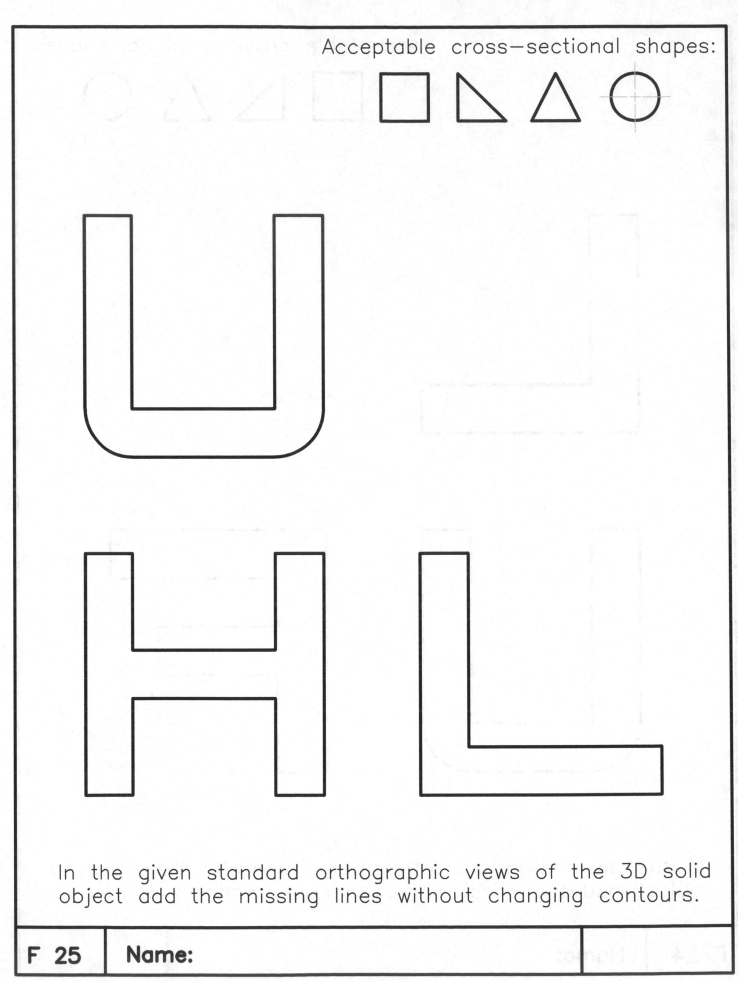

Acceptable cross-sectional shapes:

In the given standard orthographic views of the 3D solid object add the missing lines without changing contours.

F 25   Name:

In the given standard orthographic views of the 3D solid object add the missing lines without changing contours.

| F 26 | Name: | |

Chapter 3 - Second Level (S)

Level of difficulty: **Medium**

In the given standard orthographic views of the 3D solid object add the missing lines without changing contours.

In the given standard orthographic views of the 3D solid object add the missing lines without changing contours.

Name:

In the given Front, R. Side, and Bottom orthographic views of the 3D solid object add the missing lines without changing contours.

In the given Top, Front, and L. Side orthographic views of the
3D solid object add the missing lines without changing contours.

In the given standard orthographic views of the 3D solid object add the missing lines without changing contours.

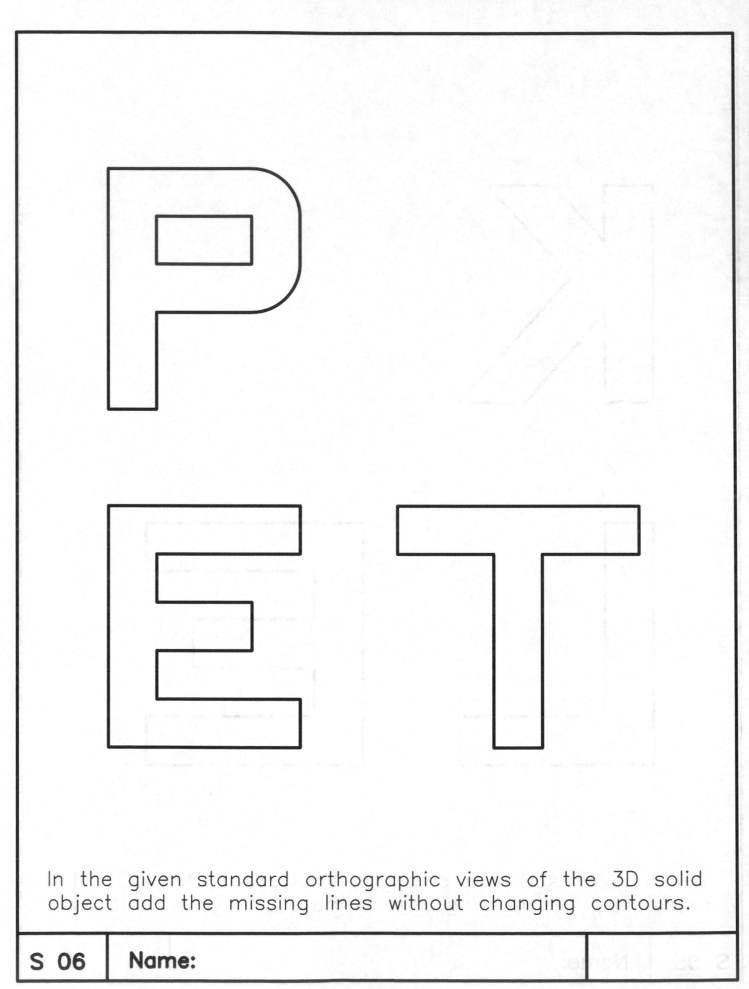

In the given standard orthographic views of the 3D solid object add the missing lines without changing contours.

Name:

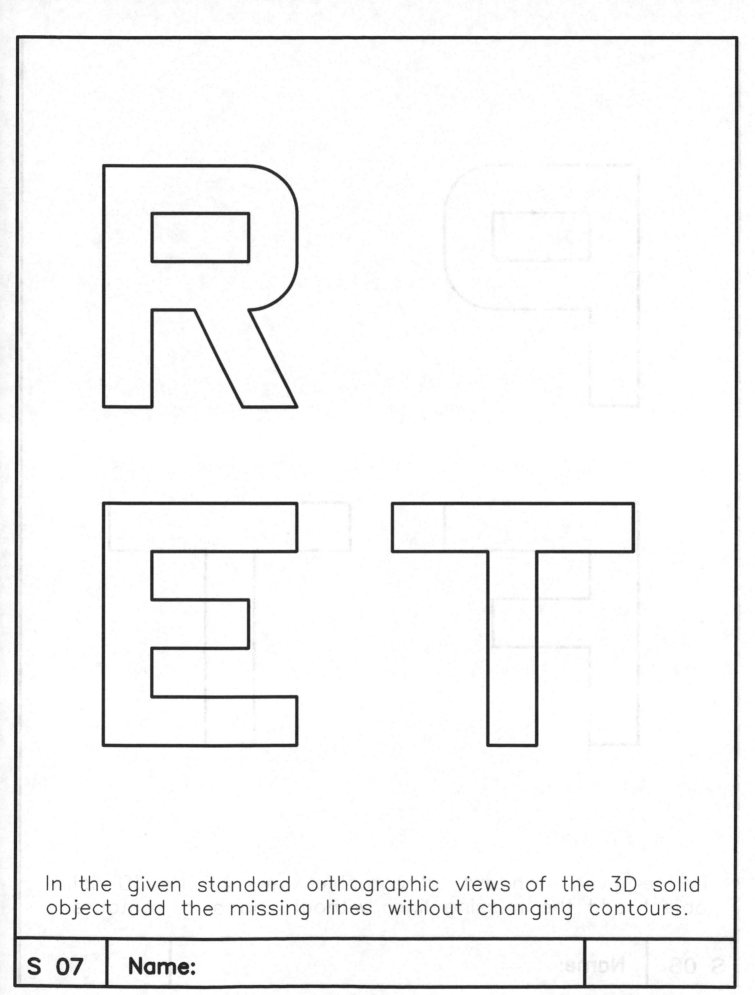

In the given standard orthographic views of the 3D solid object add the missing lines without changing contours.

Name:

In the given standard orthographic views of the 3D solid object add the missing lines without changing contours.

Name:

In the given standard orthographic views of the 3D solid object add the missing lines without changing contours.

Name:

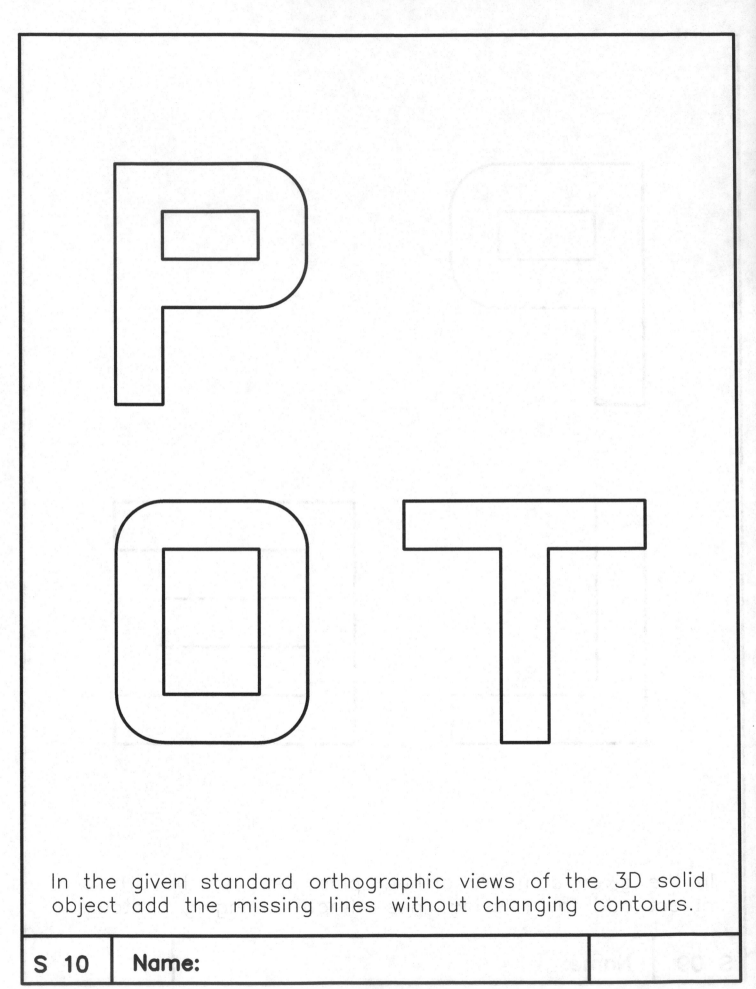

In the given standard orthographic views of the 3D solid object add the missing lines without changing contours.

Name:

In the given standard orthographic views of the 3D solid object add the missing lines without changing contours.

S 11   Name:

In the given standard orthographic views of the 3D solid object add the missing lines without changing contours.

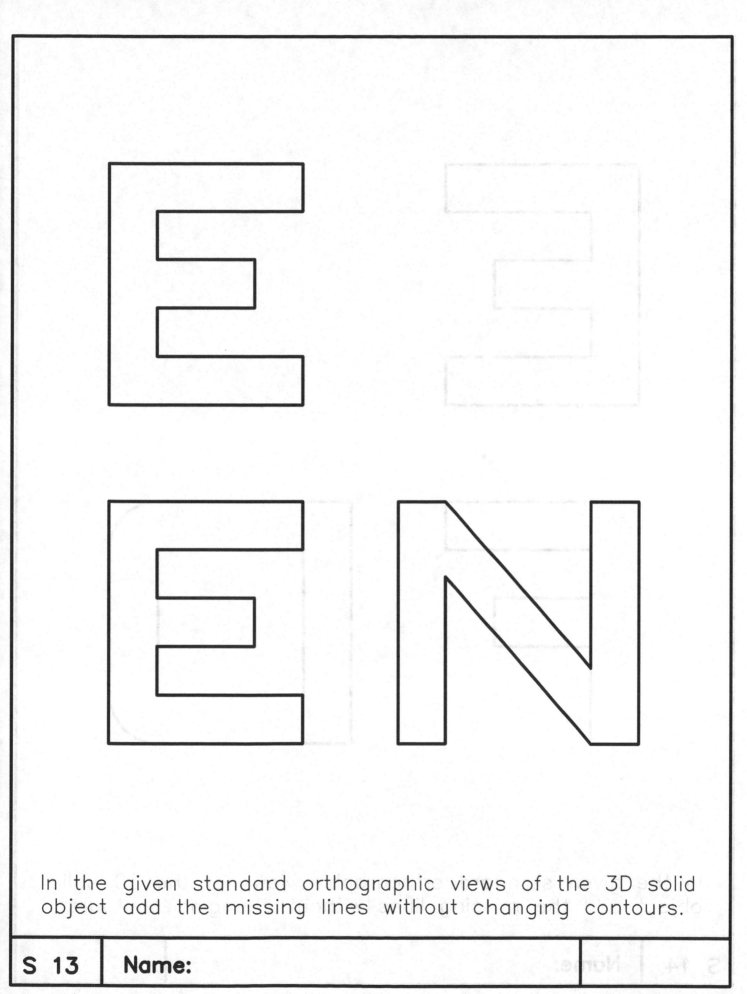

In the given standard orthographic views of the 3D solid object add the missing lines without changing contours.

Name:

In the given standard orthographic views of the 3D solid object add the missing lines without changing contours.

S 14 | Name:

In the given standard orthographic views of the 3D solid object add the missing lines without changing contours.

In the given standard orthographic views of the 3D solid object add the missing lines without changing contours.

S 16 | Name:

In the given standard orthographic views of the 3D solid object add the missing lines without changing contours.

In the given standard orthographic views of the 3D solid object add the missing lines without changing contours.

S 17b   Name:

In the given standard orthographic views of the 3D solid object add the missing lines without changing contours.

In the given standard orthographic views of the 3D solid
object add the missing lines without changing contours.

In the given standard orthographic views of the 3D solid object add the missing lines without changing contours.

Name:

In the given standard orthographic views of the 3D solid object add the missing lines without changing contours.

Name:

In the given standard orthographic views of the 3D solid object add the missing lines without changing contours.

In the given standard orthographic views of the 3D solid object add the missing lines without changing contours.

In the given standard orthographic views of the 3D solid object add the missing lines without changing contours.

Name:

In the given standard orthographic views of the 3D solid object add the missing lines without changing contours.

Name:

In the given standard orthographic views of the 3D solid object add the missing lines without changing contours.

In the given standard orthographic views of the 3D solid object add the missing lines without changing contours.

In the given standard orthographic views of the 3D solid object add the missing lines without changing contours.

S 28 | Name:

In the given Front, R. Side, and Bottom orthographic views of the 3D solid object add the missing lines without changing contours.

Name:

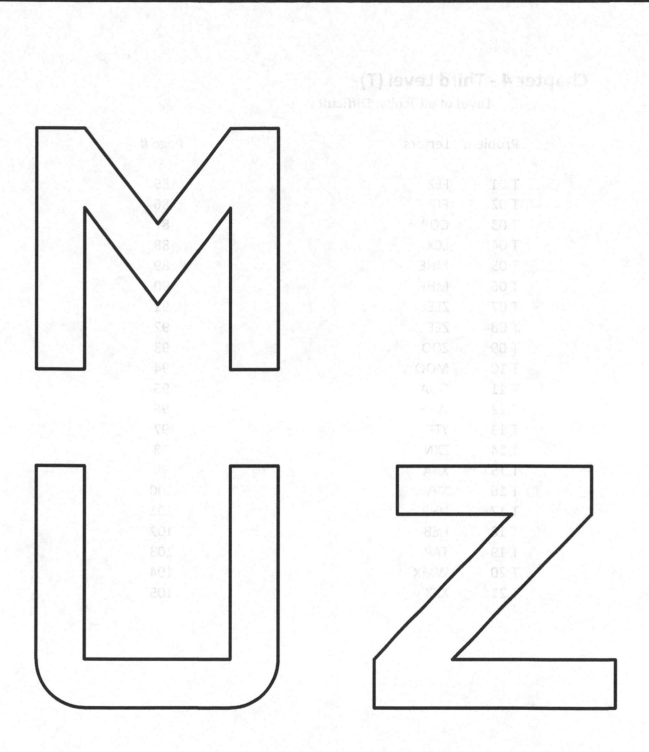

In the given standard orthographic views of the 3D solid object add the missing lines without changing contours.

Chapter 4 - Third Level (T)

Level of difficulty: **Difficult**

In the given standard orthographic views of the 3D solid object add the missing lines without changing contours.

Name:

In the given standard orthographic views of the 3D solid object add the missing lines without changing contours.

Name:

In the given standard orthographic views of the 3D solid object add the missing lines without changing contours.

In the given standard orthographic views of the 3D solid object add the missing lines without changing contours.

In the given standard orthographic views of the 3D solid object add the missing lines without changing contours.

In the given standard orthographic views of the 3D solid object add the missing lines without changing contours.

In the given standard orthographic views of the 3D solid object add the missing lines without changing contours.

In the given standard orthographic views of the 3D solid object add the missing lines without changing contours.

Name:

In the given standard orthographic views of the 3D solid object add the missing lines without changing contours.

In the given standard orthographic views of the 3D solid object add the missing lines without changing contours.

T 10 | Name:

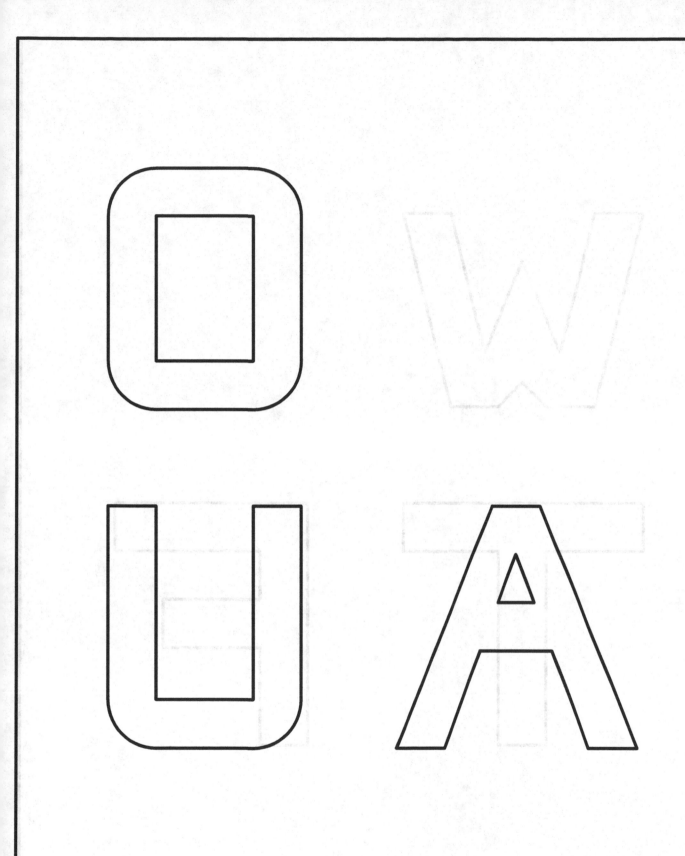

In the given standard orthographic views of the 3D solid object add the missing lines without changing contours.

In the given standard orthographic views of the 3D solid object add the missing lines without changing contours.

| Name:

In the given standard orthographic views of the 3D solid object add the missing lines without changing contours.

In the given standard orthographic views of the 3D solid object add the missing lines without changing contours.

T 14   Name:

In the given standard orthographic views of the 3D solid object add the missing lines without changing contours.

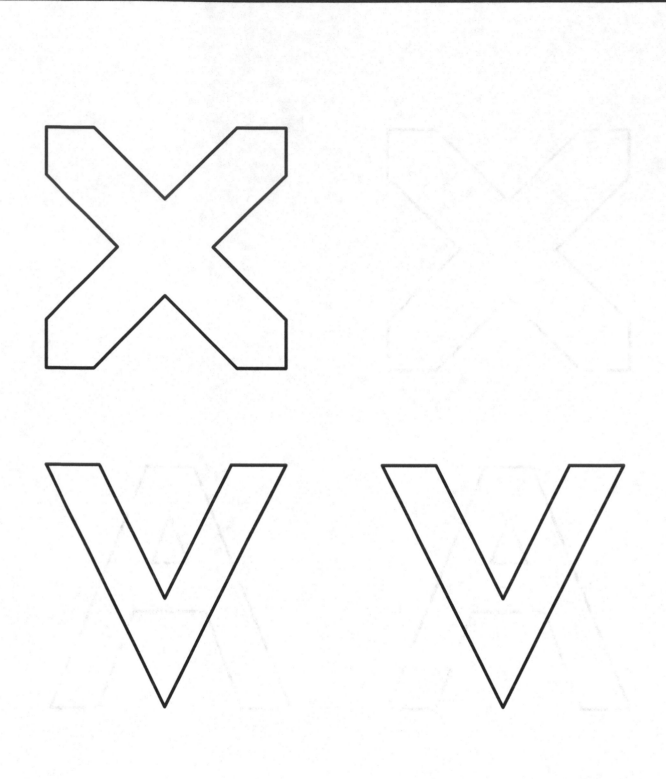

In the given standard orthographic views of the 3D solid object add the missing lines without changing contours.

Name:

In the given standard orthographic views of the 3D solid object add the missing lines without changing contours.

Name:

In the given standard orthographic views of the 3D solid object add the missing lines without changing contours.

In the given Top, Front, and L. Side orthographic views of the
3D solid object add the missing lines without changing contours.

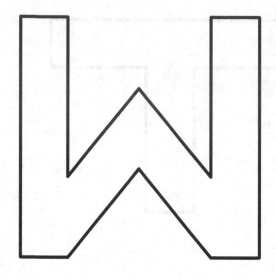

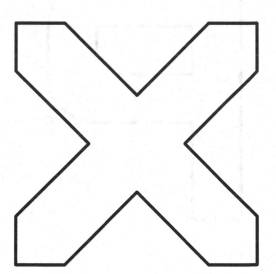

In the given standard orthographic views of the 3D solid object add the missing lines without changing contours.

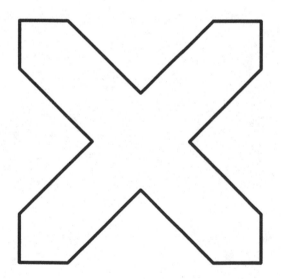

In the given standard orthographic views of the 3D solid object add the missing lines without changing contours.

List of Problem Solutions by Chapter

Chapter 1 (C) - Solutions

Top View

How many cubes make the object shown here in three views?

__9__ cubes

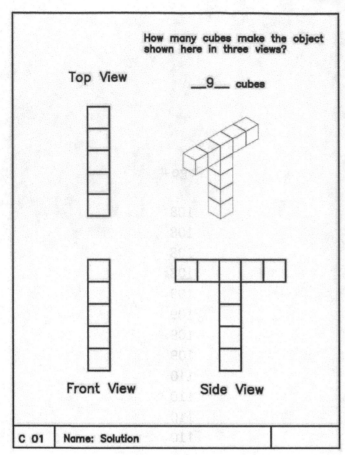

Front View   **Side View**

Top View

How many cubes make the object shown here in three views?

__9__ cubes

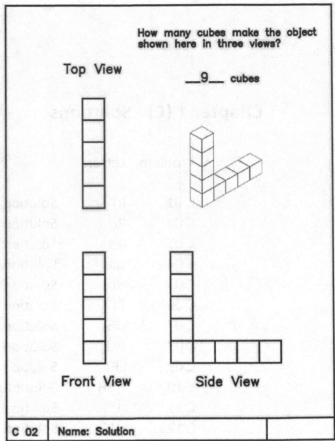

Front View   **Side View**

How many cubes make the object shown here in three views?

Top View

__13__ cubes

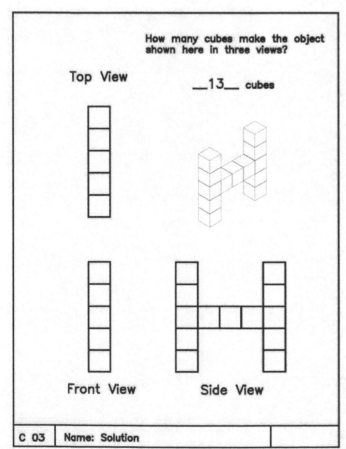

Front View   **Side View**

Top View

How many cubes make the object shown here in three views?

__13__ cubes

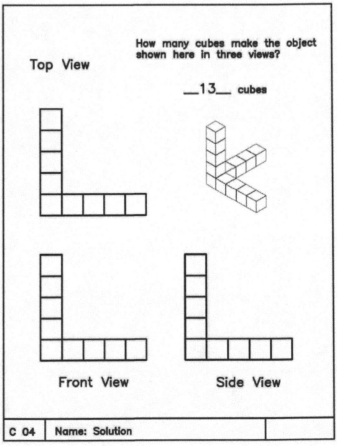

Front View   **Side View**

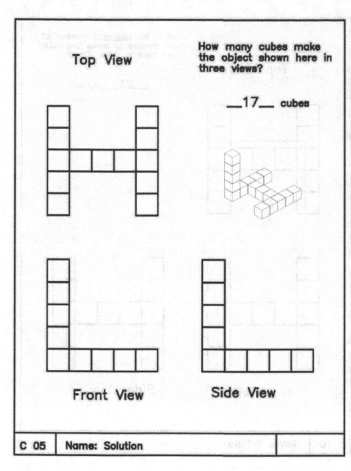

Top View

How many cubes make the object shown here in three views?

___17___ cubes

Front View

Side View

C 05   Name: Solution

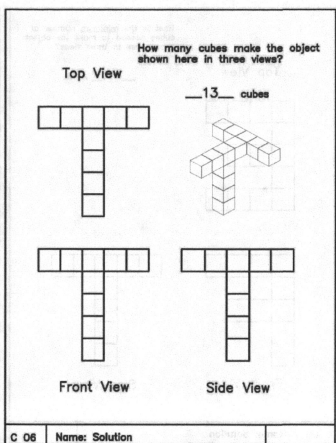

How many cubes make the object shown here in three views?

Top View

___13___ cubes

Front View

Side View

C 06   Name: Solution

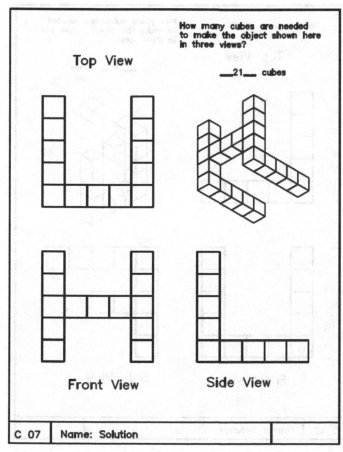

How many cubes are needed to make the object shown here in three views?

Top View

___21___ cubes

Front View

Side View

C 07   Name: Solution

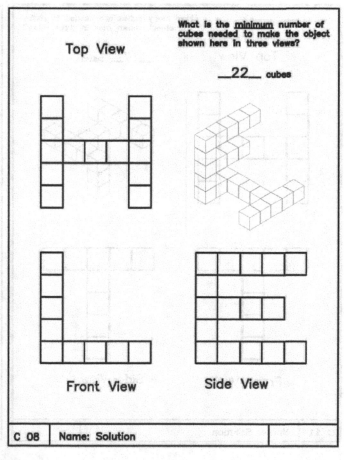

Top View

What is the **minimum** number of cubes needed to make the object shown here in three views?

___22___ cubes

Front View

Side View

C 08   Name: Solution

Top View

__17__ cubes

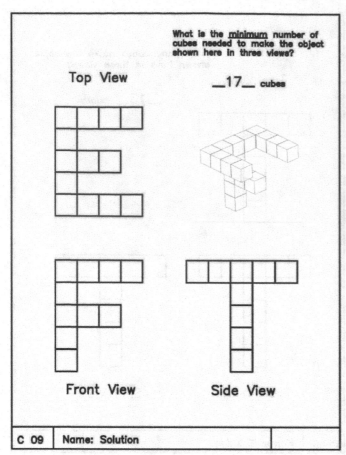

Front View　**Side View**

Top View

__21__ cubes

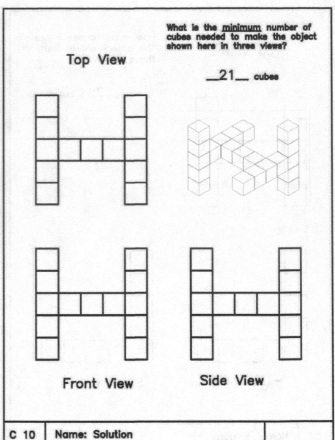

Front View　**Side View**

Top View

__17__ cubes

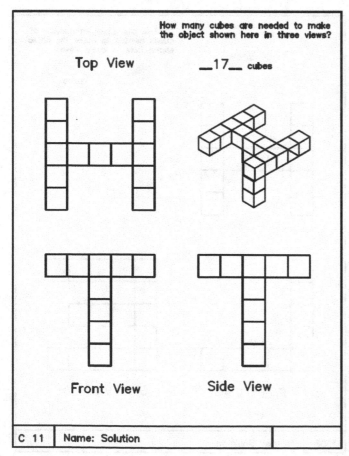

Front View　**Side View**

Top View

__19__ cubes

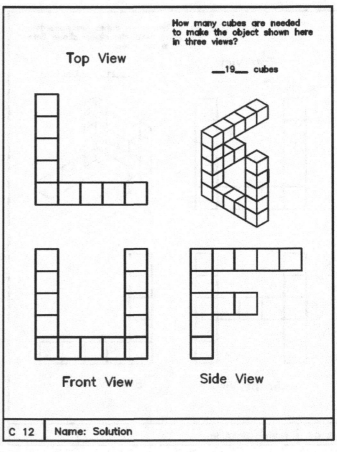

Front View　**Side View**

Chapter 2 (F) - Solutions

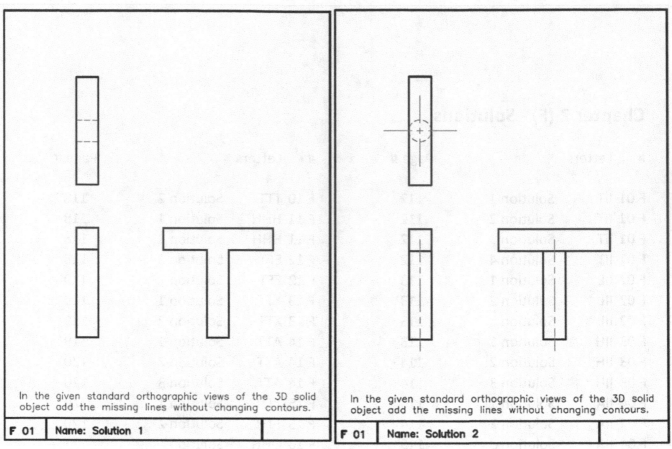

In the given standard orthographic views of the 3D solid object add the missing lines without changing contours.

In the given standard orthographic views of the 3D solid object add the missing lines without changing contours.

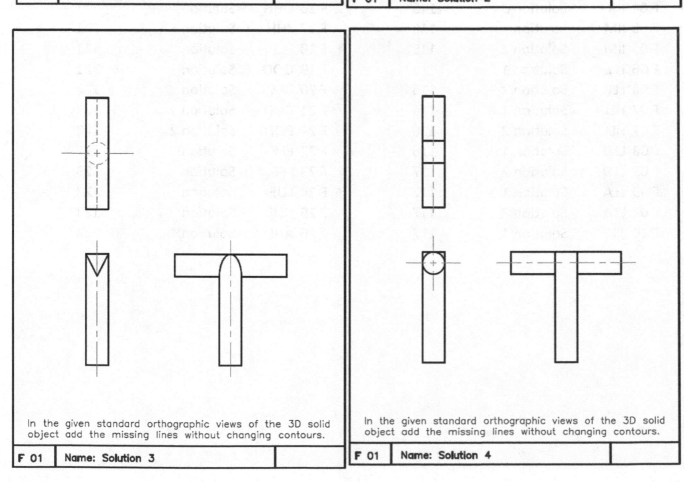

In the given standard orthographic views of the 3D solid object add the missing lines without changing contours.

In the given standard orthographic views of the 3D solid object add the missing lines without changing contours.

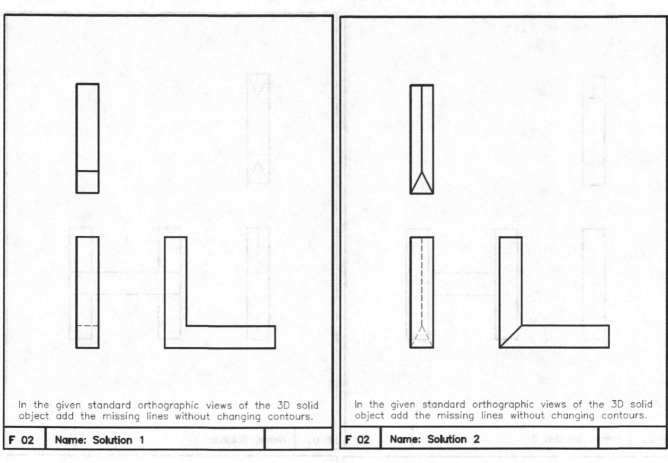

In the given standard orthographic views of the 3D solid object add the missing lines without changing contours.

| F 02 | Name: Solution 1 | |

In the given standard orthographic views of the 3D solid object add the missing lines without changing contours.

| F 02 | Name: Solution 2 | |

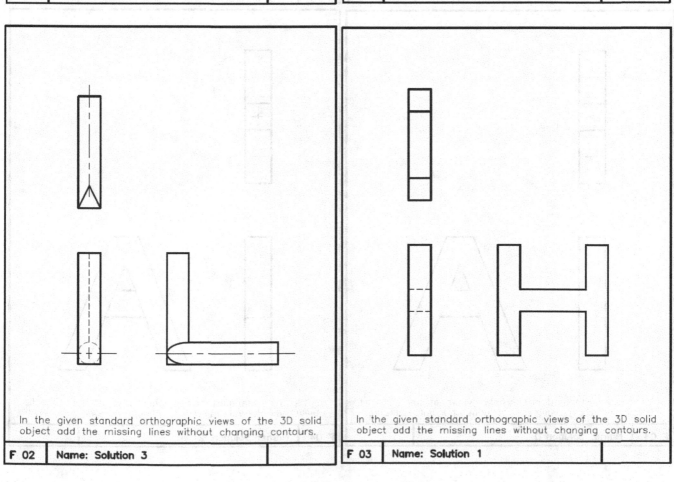

In the given standard orthographic views of the 3D solid object add the missing lines without changing contours.

| F 02 | Name: Solution 3 | |

In the given standard orthographic views of the 3D solid object add the missing lines without changing contours.

| F 03 | Name: Solution 1 | |

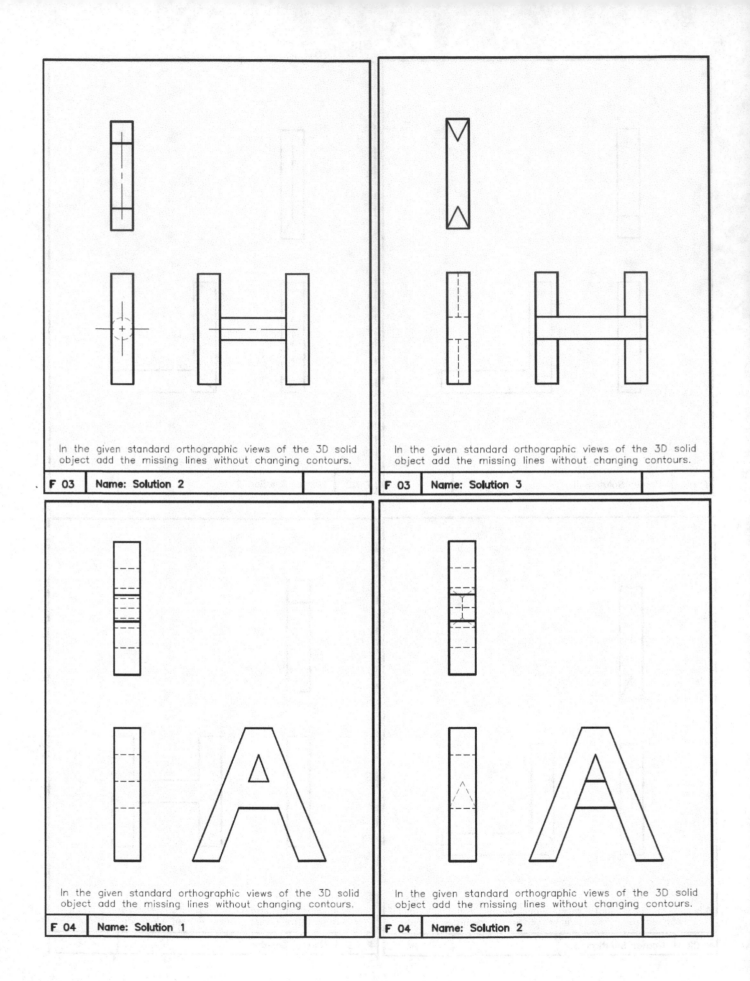

In the given standard orthographic views of the 3D solid object add the missing lines without changing contours.

F 03 | Name: Solution 2

In the given standard orthographic views of the 3D solid object add the missing lines without changing contours.

F 03 | Name: Solution 3

In the given standard orthographic views of the 3D solid object add the missing lines without changing contours.

F 04 | Name: Solution 1

In the given standard orthographic views of the 3D solid object add the missing lines without changing contours.

F 04 | Name: Solution 2

In the given standard orthographic views of the 3D solid object add the missing lines without changing contours.

| F 04 | Name: Solution 3 | |

In the given standard orthographic views of the 3D solid object add the missing lines without changing contours.

| F 05 | Name: Solution 1 | |

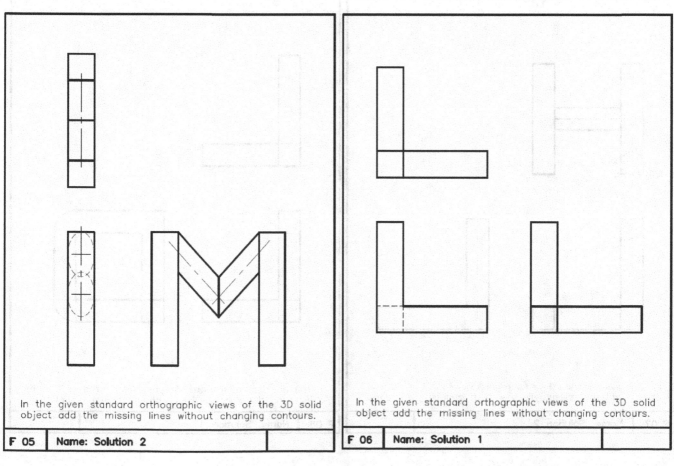

In the given standard orthographic views of the 3D solid object add the missing lines without changing contours.

| F 05 | Name: Solution 2 | |

In the given standard orthographic views of the 3D solid object add the missing lines without changing contours.

| F 06 | Name: Solution 1 | |

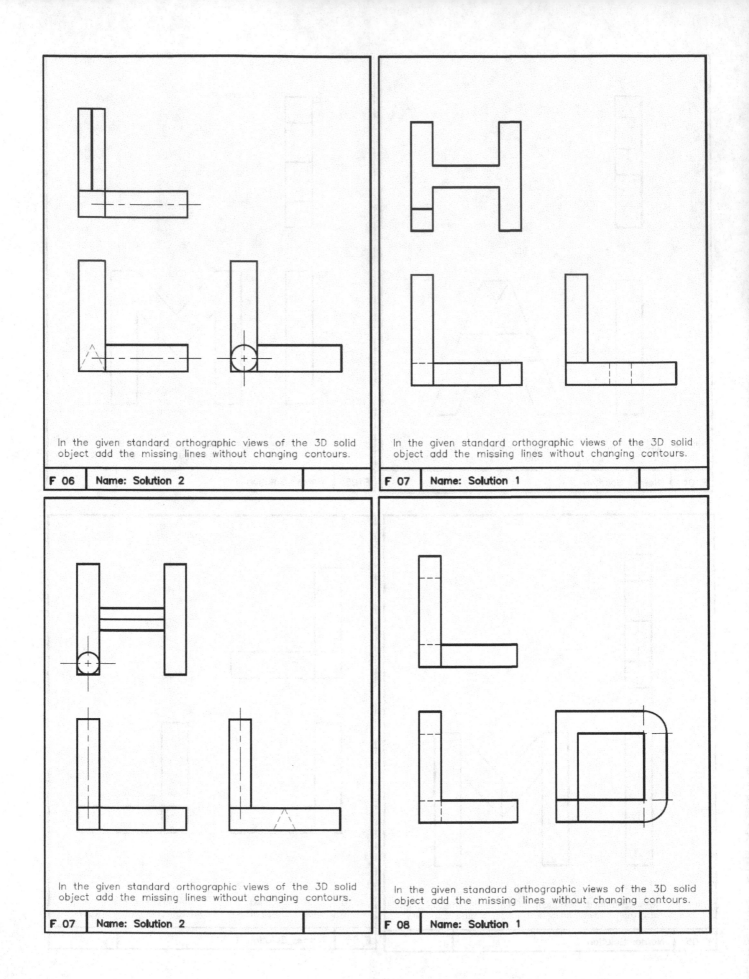

In the given standard orthographic views of the 3D solid object add the missing lines without changing contours.

| F 06 | Name: Solution 2 | |

In the given standard orthographic views of the 3D solid object add the missing lines without changing contours.

| F 07 | Name: Solution 1 | |

In the given standard orthographic views of the 3D solid object add the missing lines without changing contours.

| F 07 | Name: Solution 2 | |

In the given standard orthographic views of the 3D solid object add the missing lines without changing contours.

| F 08 | Name: Solution 1 | |

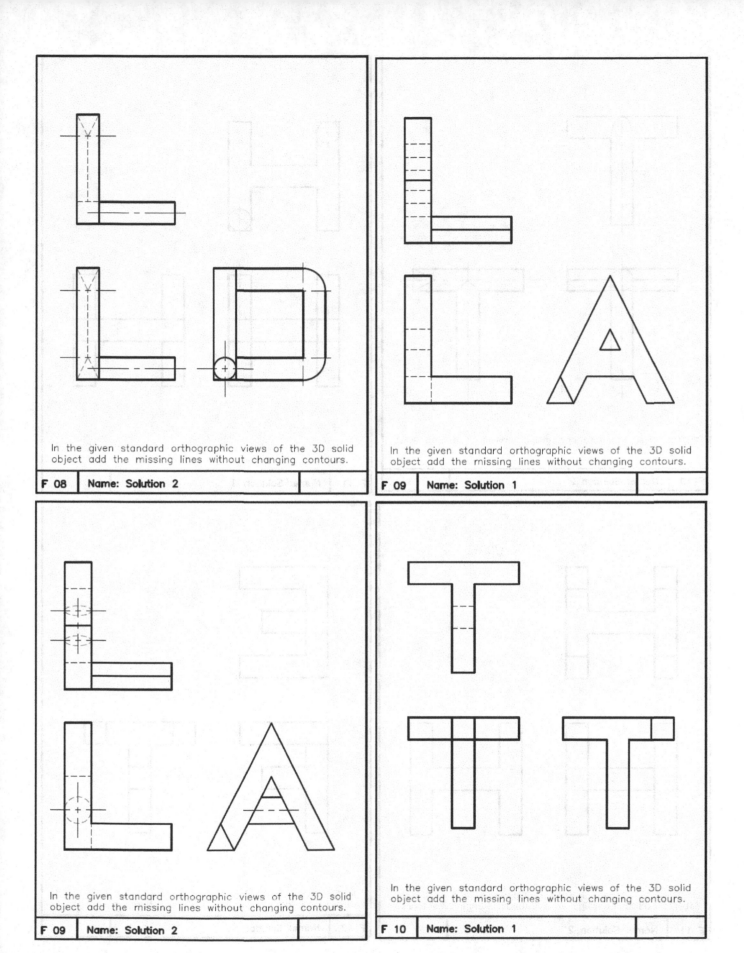

In the given standard orthographic views of the 3D solid object add the missing lines without changing contours.

F 08 | Name: Solution 2

In the given standard orthographic views of the 3D solid object add the missing lines without changing contours.

F 09 | Name: Solution 1

In the given standard orthographic views of the 3D solid object add the missing lines without changing contours.

F 09 | Name: Solution 2

In the given standard orthographic views of the 3D solid object add the missing lines without changing contours.

F 10 | Name: Solution 1

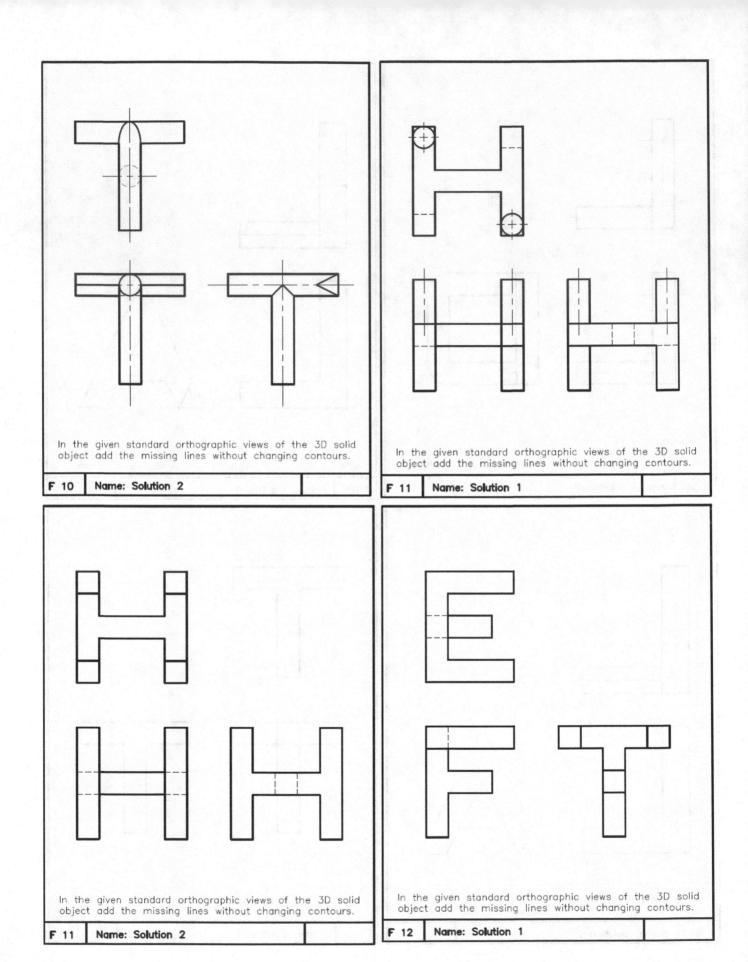

In the given standard orthographic views of the 3D solid object add the missing lines without changing contours.

F 10 | Name: Solution 2

In the given standard orthographic views of the 3D solid object add the missing lines without changing contours.

F 11 | Name: Solution 1

In the given standard orthographic views of the 3D solid object add the missing lines without changing contours.

F 11 | Name: Solution 2

In the given standard orthographic views of the 3D solid object add the missing lines without changing contours.

F 12 | Name: Solution 1

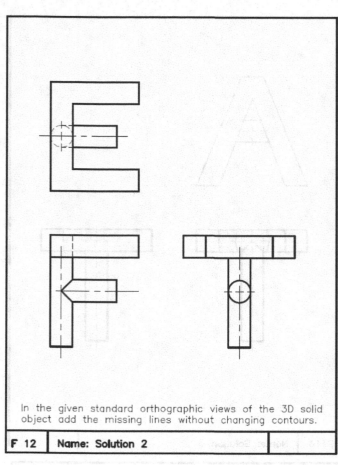

In the given standard orthographic views of the 3D solid object add the missing lines without changing contours.

F 12   Name: Solution 2

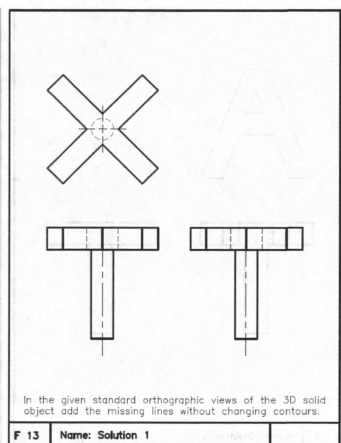

In the given standard orthographic views of the 3D solid object add the missing lines without changing contours.

F 13   Name: Solution 1

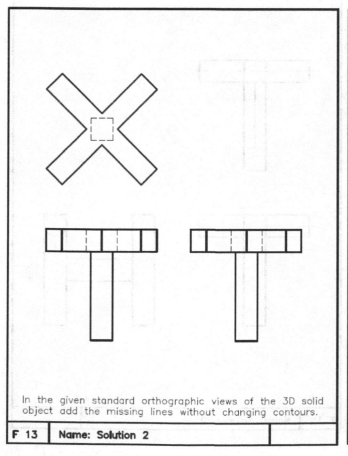

In the given standard orthographic views of the 3D solid object add the missing lines without changing contours.

F 13   Name: Solution 2

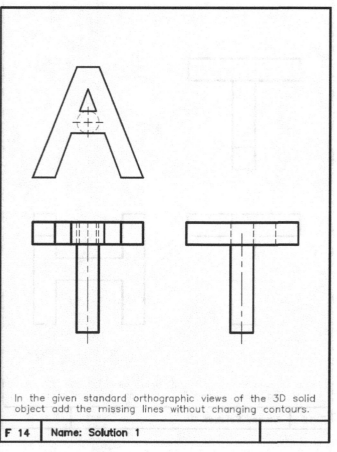

In the given standard orthographic views of the 3D solid object add the missing lines without changing contours.

F 14   Name: Solution 1

119

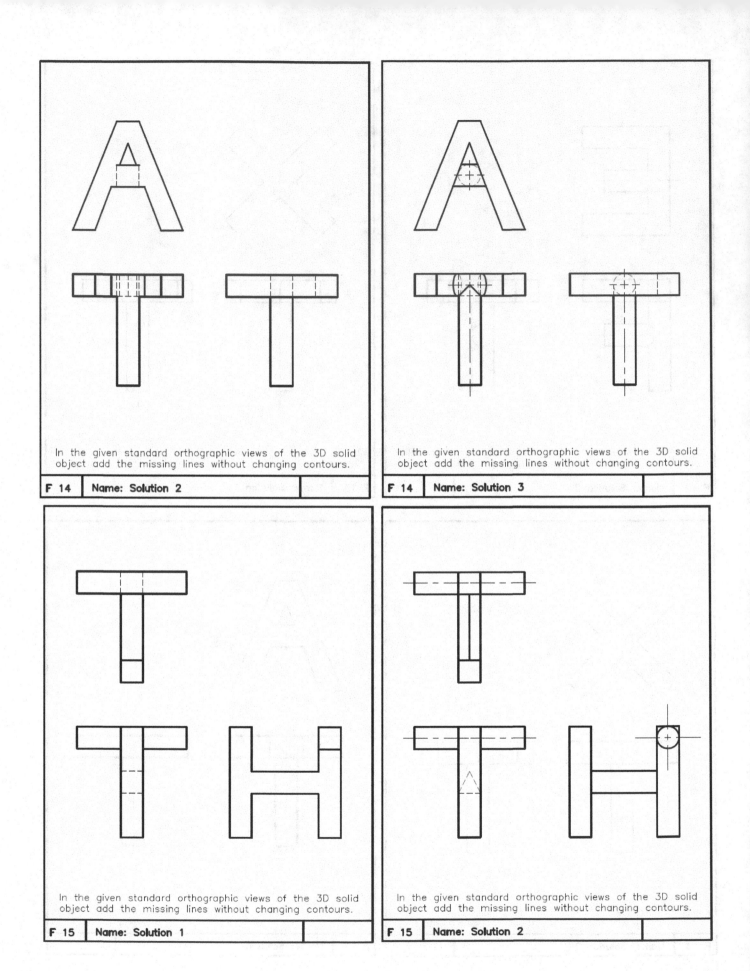

In the given standard orthographic views of the 3D solid object add the missing lines without changing contours.

| F 14 | Name: Solution 2 | |

In the given standard orthographic views of the 3D solid object add the missing lines without changing contours.

| F 14 | Name: Solution 3 | |

In the given standard orthographic views of the 3D solid object add the missing lines without changing contours.

| F 15 | Name: Solution 1 | |

In the given standard orthographic views of the 3D solid object add the missing lines without changing contours.

| F 15 | Name: Solution 2 | |

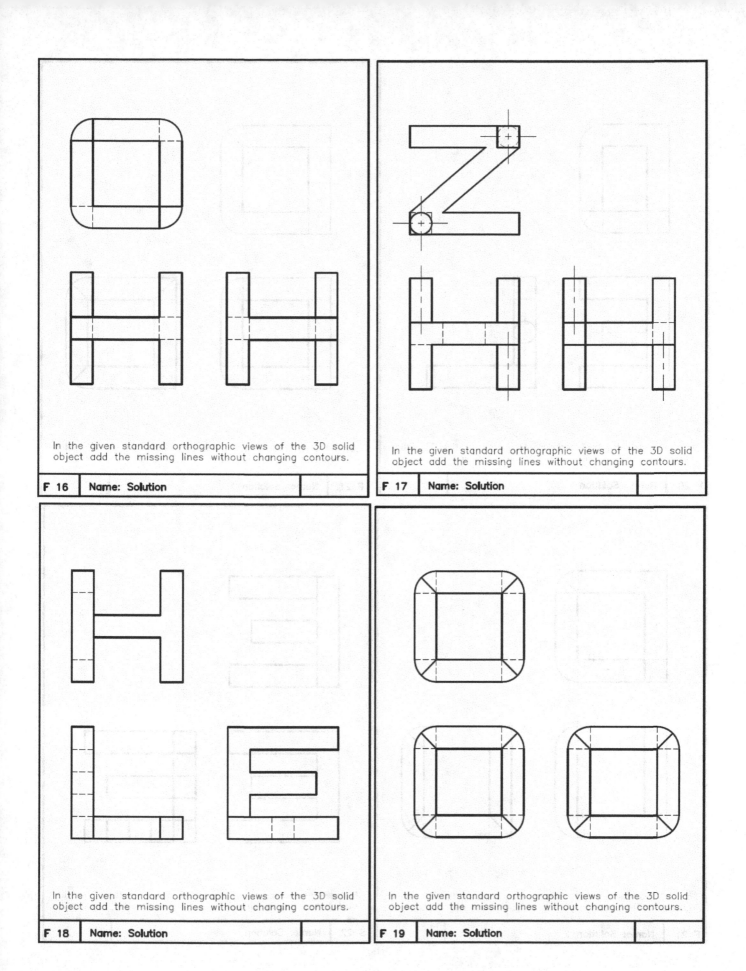

In the given standard orthographic views of the 3D solid object add the missing lines without changing contours.

F 16   Name: Solution

In the given standard orthographic views of the 3D solid object add the missing lines without changing contours.

F 17   Name: Solution

In the given standard orthographic views of the 3D solid object add the missing lines without changing contours.

F 18   Name: Solution

In the given standard orthographic views of the 3D solid object add the missing lines without changing contours.

F 19   Name: Solution

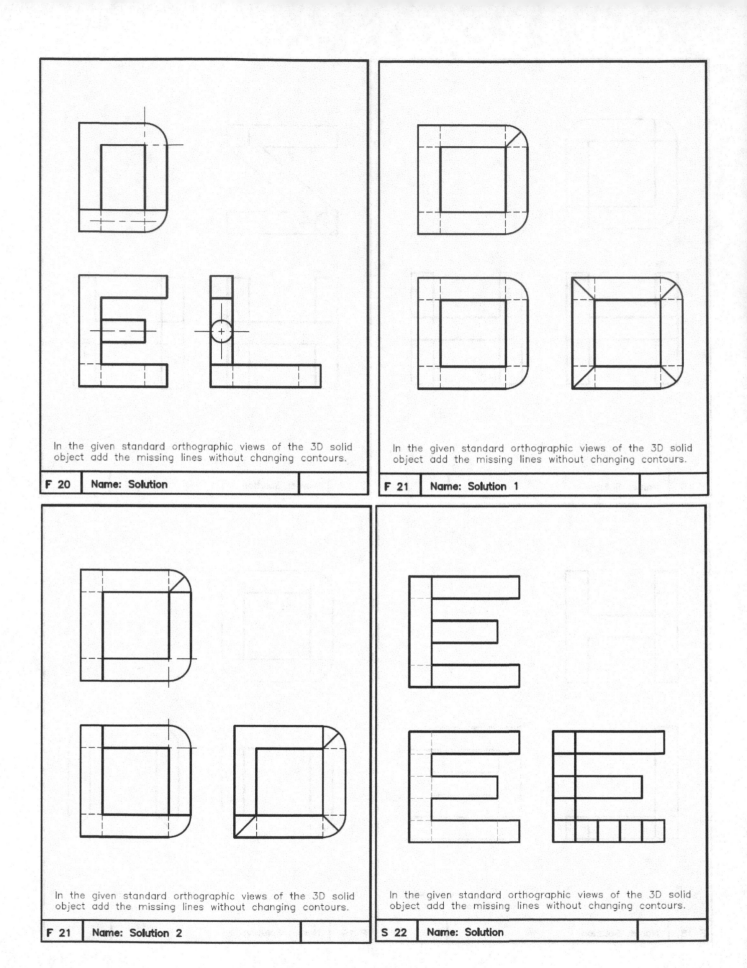

In the given standard orthographic views of the 3D solid object add the missing lines without changing contours.

| F 20 | Name: Solution |

In the given standard orthographic views of the 3D solid object add the missing lines without changing contours.

| F 21 | Name: Solution 1 |

In the given standard orthographic views of the 3D solid object add the missing lines without changing contours.

| F 21 | Name: Solution 2 |

In the given standard orthographic views of the 3D solid object add the missing lines without changing contours.

| S 22 | Name: Solution |

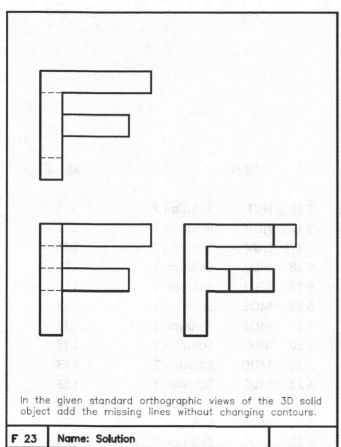

In the given standard orthographic views of the 3D solid object add the missing lines without changing contours.

| F 23 | Name: Solution | |

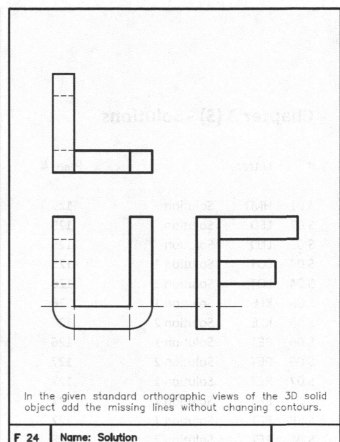

In the given standard orthographic views of the 3D solid object add the missing lines without changing contours.

| F 24 | Name: Solution | |

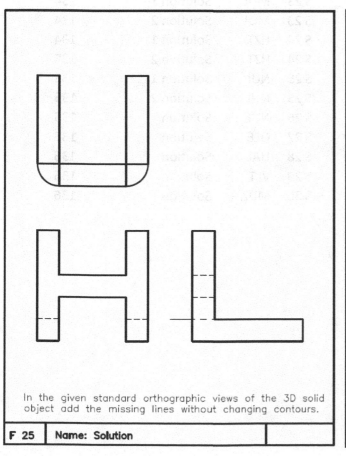

In the given standard orthographic views of the 3D solid object add the missing lines without changing contours.

| F 25 | Name: Solution | |

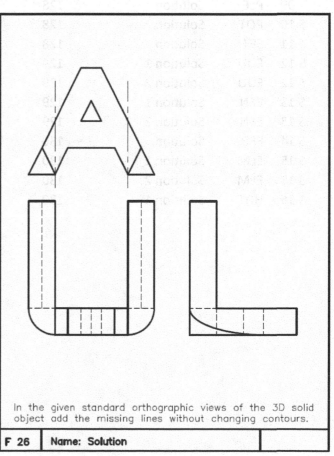

In the given standard orthographic views of the 3D solid object add the missing lines without changing contours.

| F 26 | Name: Solution | |

Chapter 3 (S) - Solutions

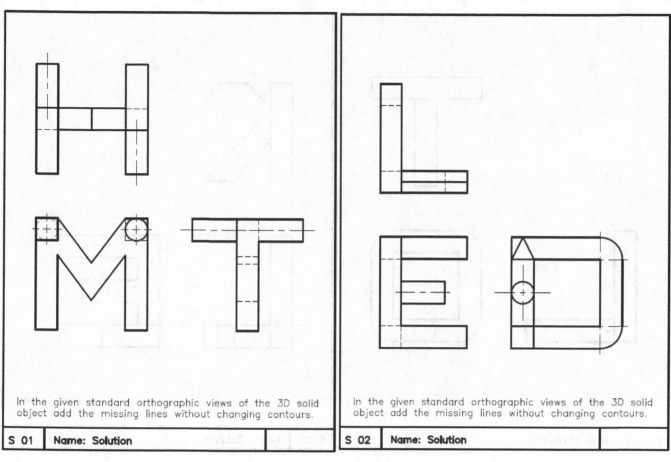

In the given standard orthographic views of the 3D solid object add the missing lines without changing contours.

S 01 | Name: Solution

In the given standard orthographic views of the 3D solid object add the missing lines without changing contours.

S 02 | Name: Solution

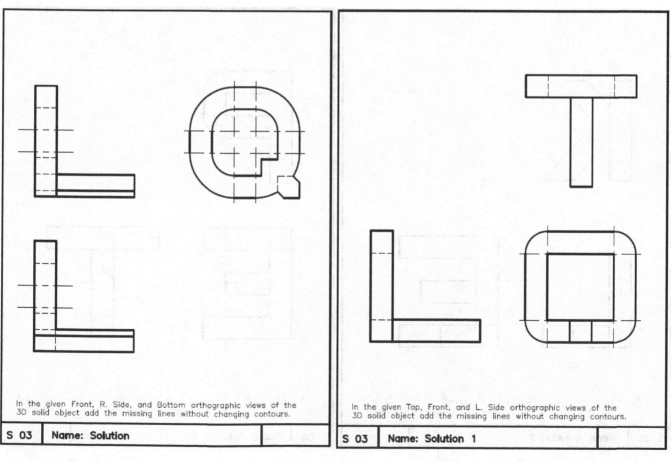

In the given Front, R. Side, and Bottom orthographic views of the 3D solid object add the missing lines without changing contours.

S 03 | Name: Solution

In the given Top, Front, and L. Side orthographic views of the 3D solid object add the missing lines without changing contours.

S 03 | Name: Solution 1

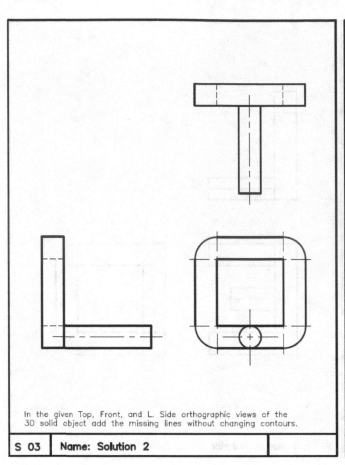

In the given Top, Front, and L. Side orthographic views of the 3D solid object add the missing lines without changing contours.

S 03 | Name: Solution 2

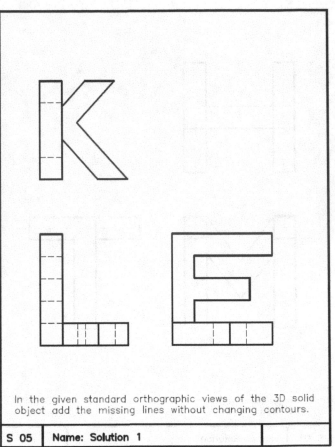

In the given standard orthographic views of the 3D solid object add the missing lines without changing contours.

S 05 | Name: Solution 1

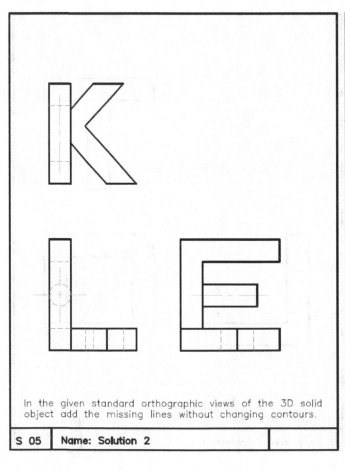

In the given standard orthographic views of the 3D solid object add the missing lines without changing contours.

S 05 | Name: Solution 2

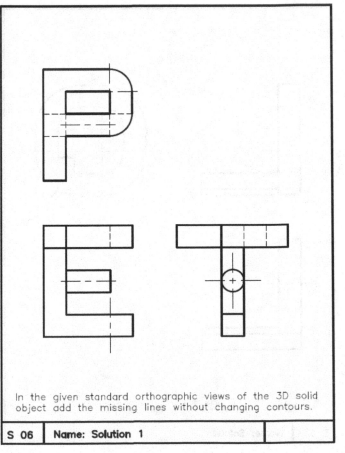

In the given standard orthographic views of the 3D solid object add the missing lines without changing contours.

S 06 | Name: Solution 1

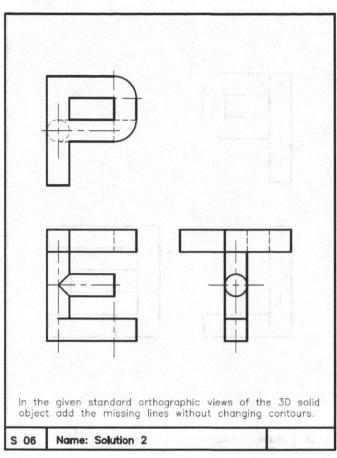

In the given standard orthographic views of the 3D solid object add the missing lines without changing contours.

| S 06 | Name: Solution 2 | |

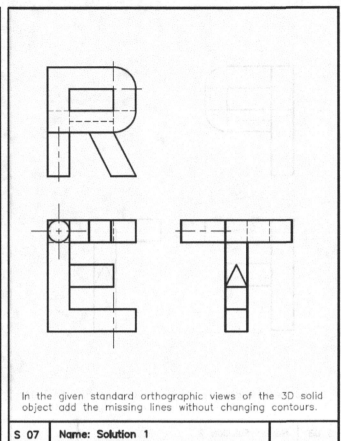

In the given standard orthographic views of the 3D solid object add the missing lines without changing contours.

| S 07 | Name: Solution 1 | |

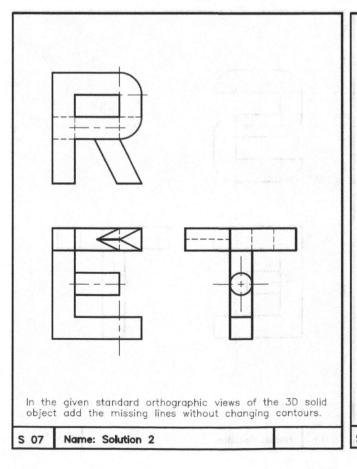

In the given standard orthographic views of the 3D solid object add the missing lines without changing contours.

| S 07 | Name: Solution 2 | |

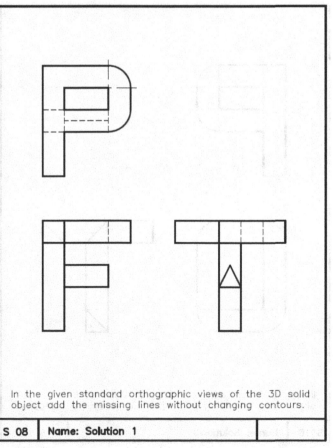

In the given standard orthographic views of the 3D solid object add the missing lines without changing contours.

| S 08 | Name: Solution 1 | |

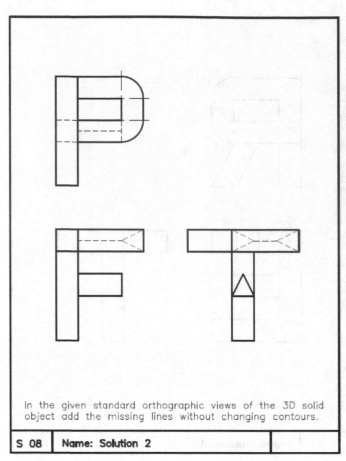

In the given standard orthographic views of the 3D solid object add the missing lines without changing contours.

| S 08 | Name: Solution 2 | |

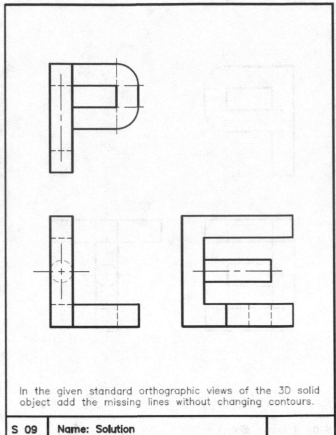

In the given standard orthographic views of the 3D solid object add the missing lines without changing contours.

| S 09 | Name: Solution | |

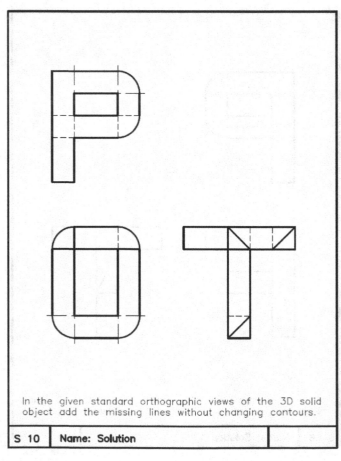

In the given standard orthographic views of the 3D solid object add the missing lines without changing contours.

| S 10 | Name: Solution | |

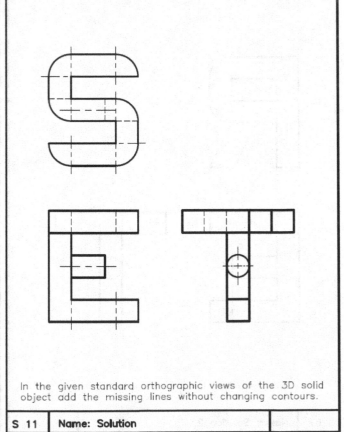

In the given standard orthographic views of the 3D solid object add the missing lines without changing contours.

| S 11 | Name: Solution | |

128

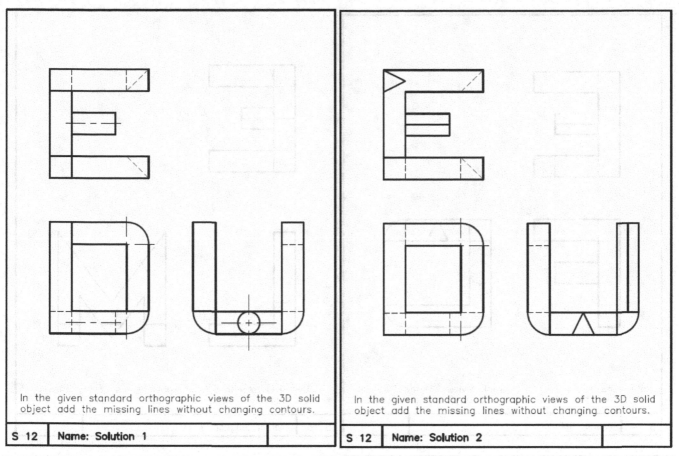

In the given standard orthographic views of the 3D solid object add the missing lines without changing contours.

| S 12 | Name: Solution 1 | |

In the given standard orthographic views of the 3D solid object add the missing lines without changing contours.

| S 12 | Name: Solution 2 | |

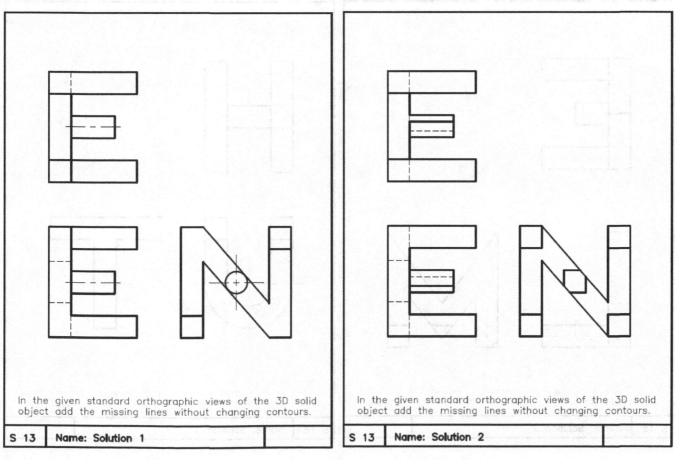

In the given standard orthographic views of the 3D solid object add the missing lines without changing contours.

| S 13 | Name: Solution 1 | |

In the given standard orthographic views of the 3D solid object add the missing lines without changing contours.

| S 13 | Name: Solution 2 | |

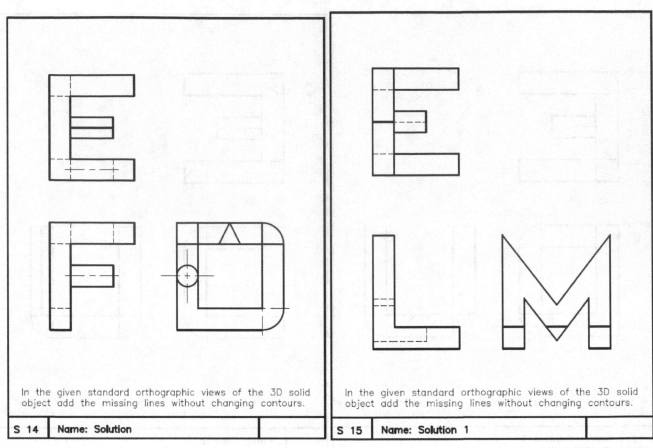

In the given standard orthographic views of the 3D solid object add the missing lines without changing contours.

| S 14 | Name: Solution | |

In the given standard orthographic views of the 3D solid object add the missing lines without changing contours.

| S 15 | Name: Solution 1 | |

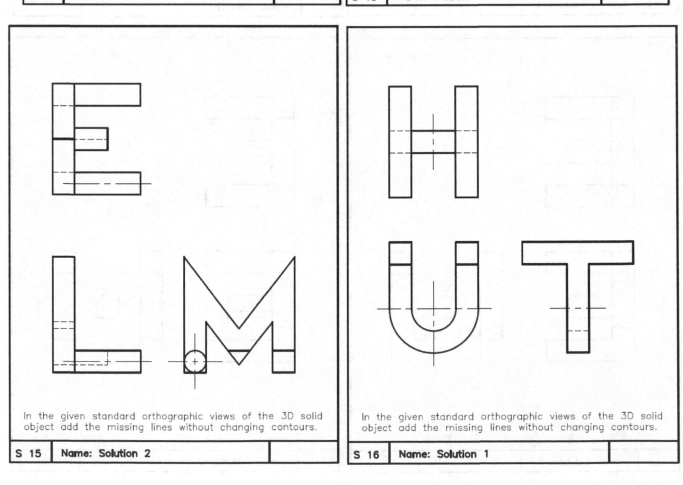

In the given standard orthographic views of the 3D solid object add the missing lines without changing contours.

| S 15 | Name: Solution 2 | |

In the given standard orthographic views of the 3D solid object add the missing lines without changing contours.

| S 16 | Name: Solution 1 | |

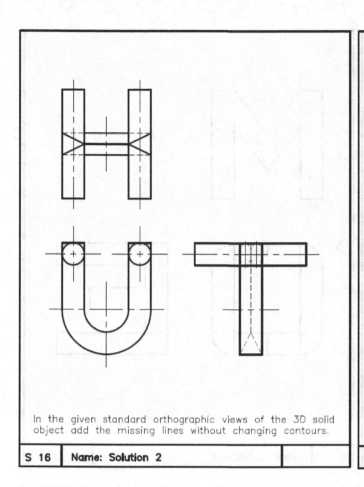

In the given standard orthographic views of the 3D solid object add the missing lines without changing contours.

| S 16 | Name: Solution 2 | |

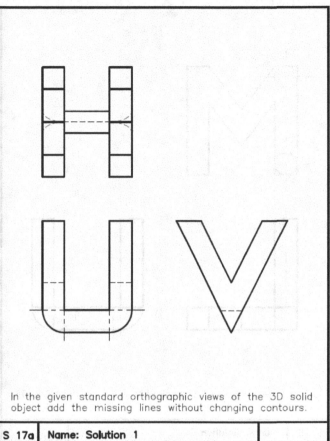

In the given standard orthographic views of the 3D solid object add the missing lines without changing contours.

| S 17a | Name: Solution 1 | |

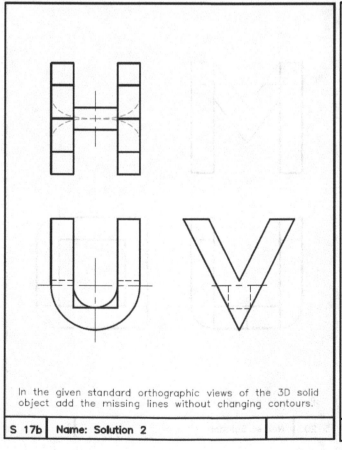

In the given standard orthographic views of the 3D solid object add the missing lines without changing contours.

| S 17b | Name: Solution 2 | |

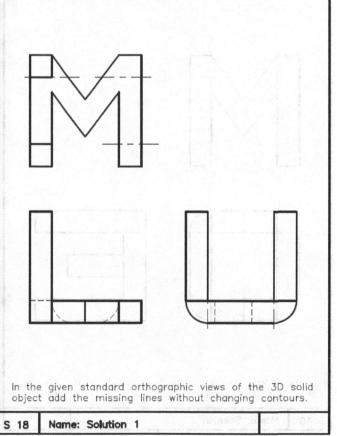

In the given standard orthographic views of the 3D solid object add the missing lines without changing contours.

| S 18 | Name: Solution 1 | |

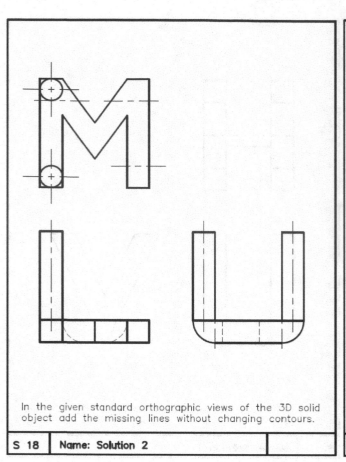

In the given standard orthographic views of the 3D solid object add the missing lines without changing contours.

S 18	Name: Solution 2

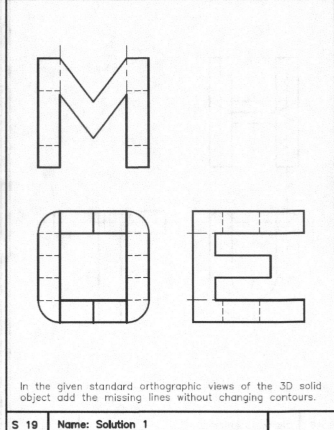

In the given standard orthographic views of the 3D solid object add the missing lines without changing contours.

S 19	Name: Solution 1

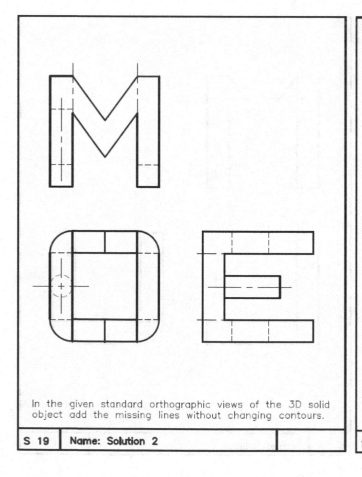

In the given standard orthographic views of the 3D solid object add the missing lines without changing contours.

S 19	Name: Solution 2

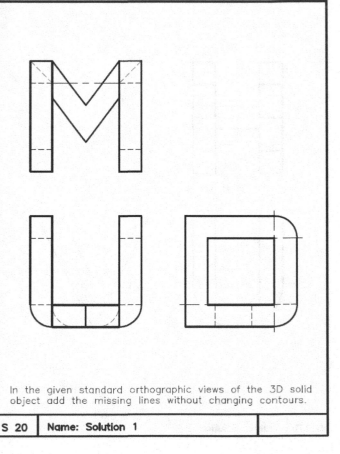

In the given standard orthographic views of the 3D solid object add the missing lines without changing contours.

S 20	Name: Solution 1

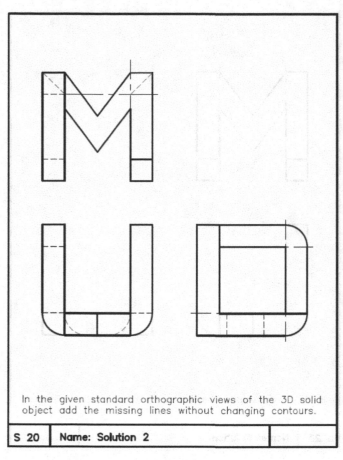

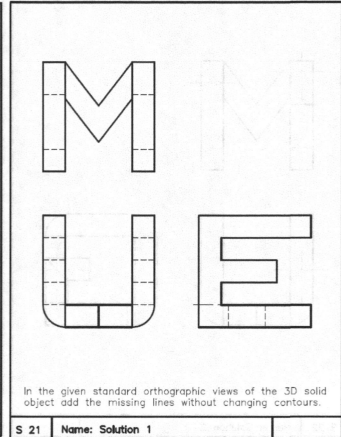

In the given standard orthographic views of the 3D solid object add the missing lines without changing contours.

| S 20 | Name: Solution 2 | | |

In the given standard orthographic views of the 3D solid object add the missing lines without changing contours.

| S 21 | Name: Solution 1 | | |

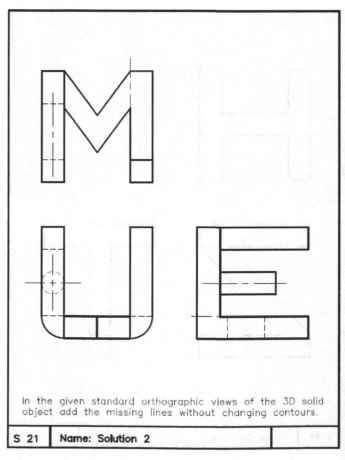

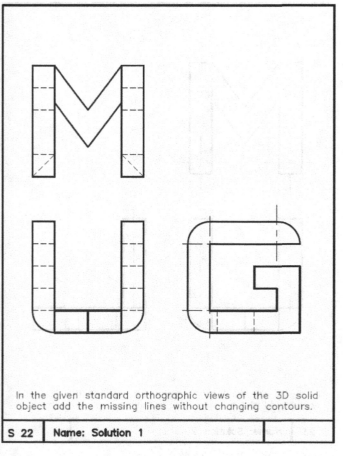

In the given standard orthographic views of the 3D solid object add the missing lines without changing contours.

| S 21 | Name: Solution 2 | | |

In the given standard orthographic views of the 3D solid object add the missing lines without changing contours.

| S 22 | Name: Solution 1 | | |

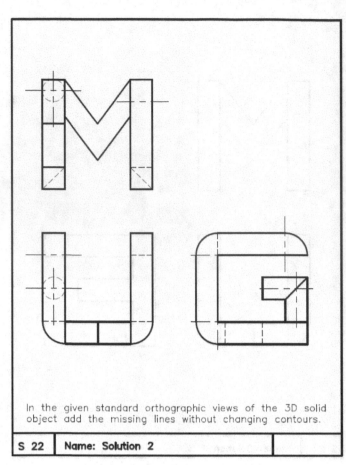

In the given standard orthographic views of the 3D solid object add the missing lines without changing contours.

| S 22 | Name: Solution 2 | |

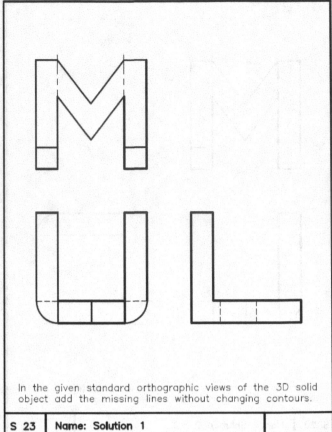

In the given standard orthographic views of the 3D solid object add the missing lines without changing contours.

| S 23 | Name: Solution 1 | |

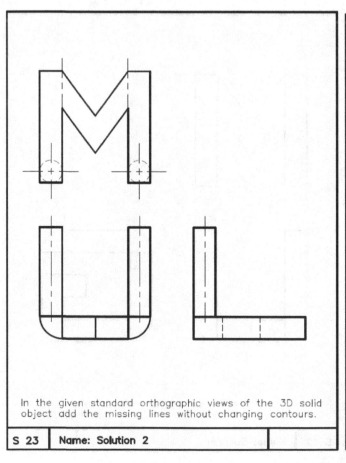

In the given standard orthographic views of the 3D solid object add the missing lines without changing contours.

| S 23 | Name: Solution 2 | |

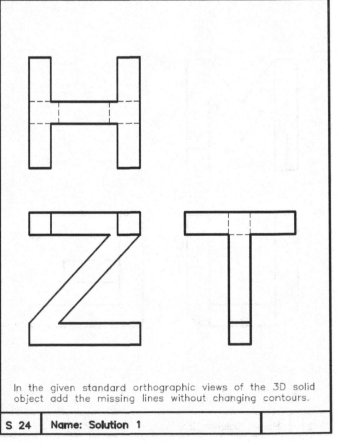

In the given standard orthographic views of the 3D solid object add the missing lines without changing contours.

| S 24 | Name: Solution 1 | |

134

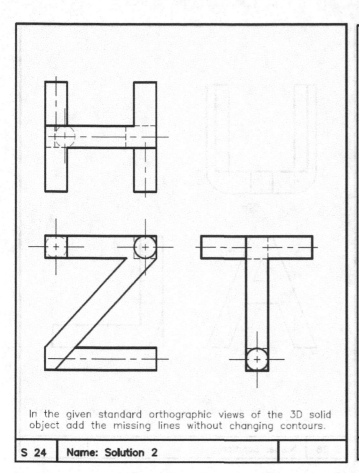

In the given standard orthographic views of the 3D solid object add the missing lines without changing contours.

| S 24 | Name: Solution 2 | |

In the given standard orthographic views of the 3D solid object add the missing lines without changing contours.

| S 25 | Name: Solution 1 | |

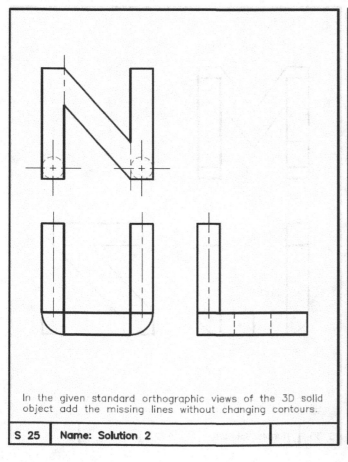

In the given standard orthographic views of the 3D solid object add the missing lines without changing contours.

| S 25 | Name: Solution 2 | |

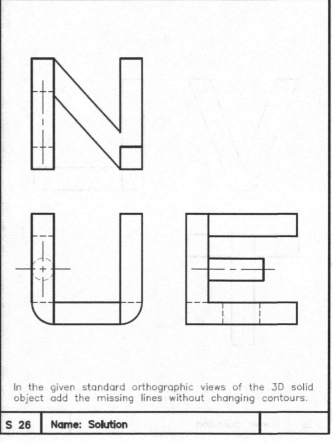

In the given standard orthographic views of the 3D solid object add the missing lines without changing contours.

| S 26 | Name: Solution | |

In the given standard orthographic views of the 3D solid object add the missing lines without changing contours.

| S 27 | Name: Solution | |

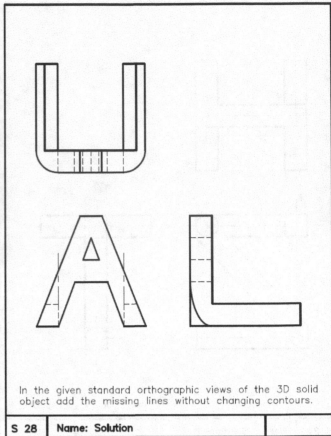

In the given standard orthographic views of the 3D solid object add the missing lines without changing contours.

| S 28 | Name: Solution | |

In the given Front, R. Side, and Bottom orthographic views of the 3D solid object add the missing lines without changing contours.

| S 29 | Name: Solutioon | |

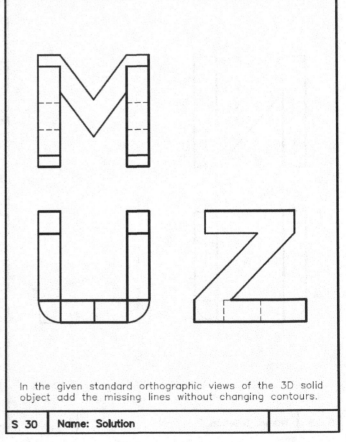

In the given standard orthographic views of the 3D solid object add the missing lines without changing contours.

| S 30 | Name: Solution | |

136

Chapter 4 (T) - Solutions

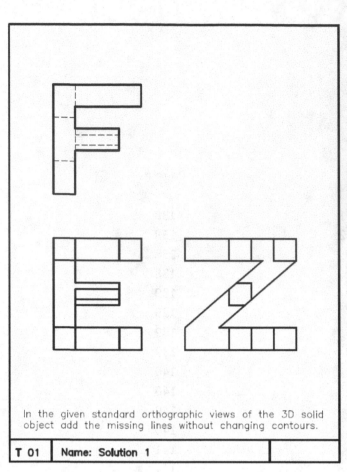

In the given standard orthographic views of the 3D solid object add the missing lines without changing contours.

| T 01 | Name: Solution 1 | |

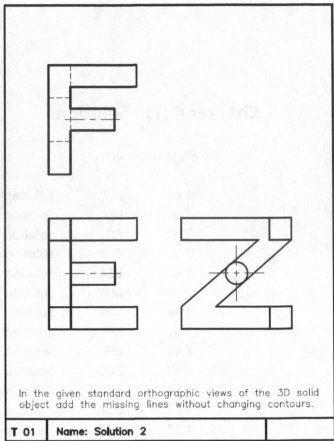

In the given standard orthographic views of the 3D solid object add the missing lines without changing contours.

| T 01 | Name: Solution 2 | |

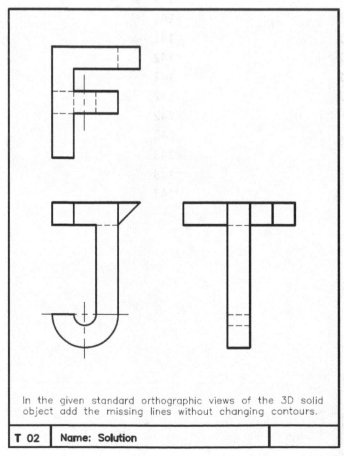

In the given standard orthographic views of the 3D solid object add the missing lines without changing contours.

| T 02 | Name: Solution | |

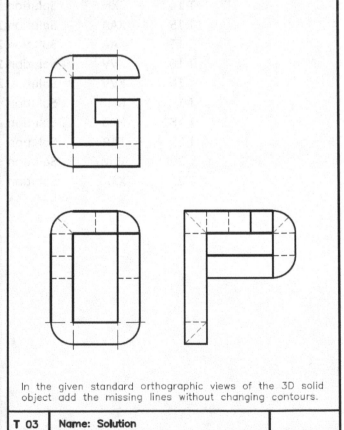

In the given standard orthographic views of the 3D solid object add the missing lines without changing contours.

| T 03 | Name: Solution | |

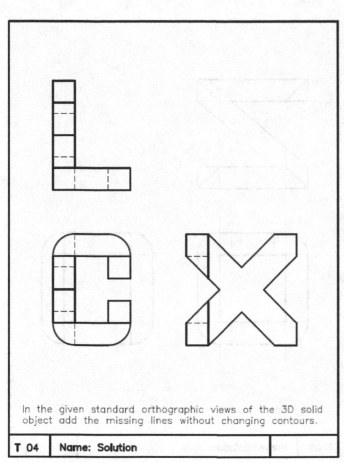

T 04 Name: Solution

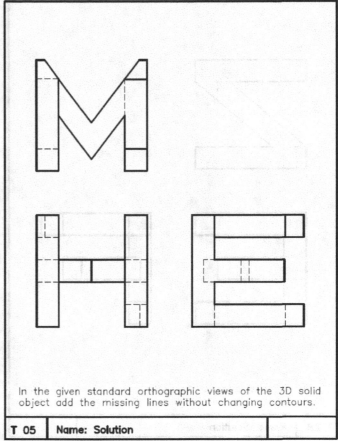

T 05 Name: Solution

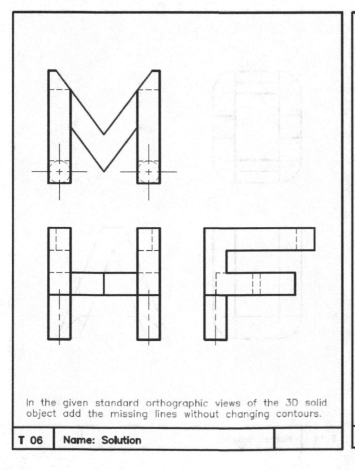

T 06 Name: Solution

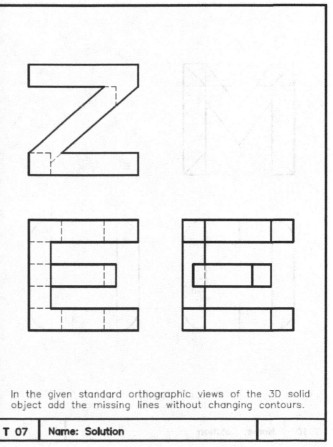

T 07 Name: Solution

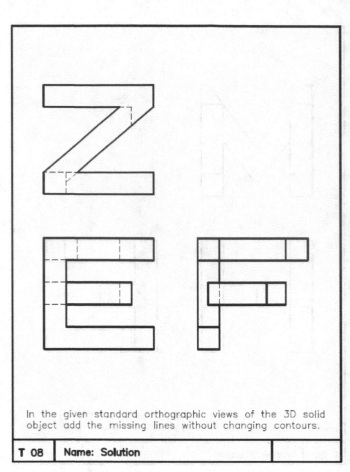

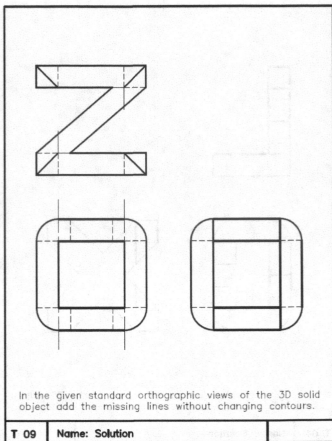

In the given standard orthographic views of the 3D solid object add the missing lines without changing contours.

| T 08 | Name: Solution | |

In the given standard orthographic views of the 3D solid object add the missing lines without changing contours.

| T 09 | Name: Solution | |

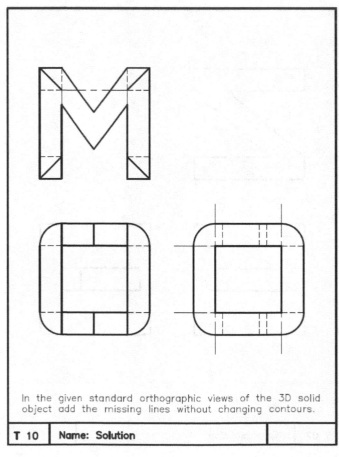

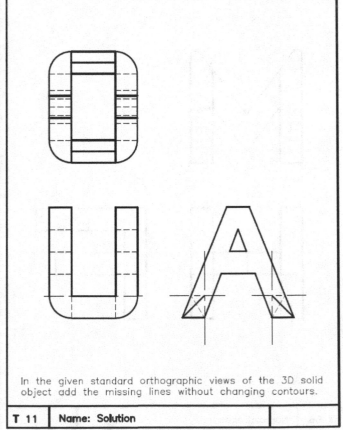

In the given standard orthographic views of the 3D solid object add the missing lines without changing contours.

| T 10 | Name: Solution | |

In the given standard orthographic views of the 3D solid object add the missing lines without changing contours.

| T 11 | Name: Solution | |

In the given standard orthographic views of the 3D solid object add the missing lines without changing contours.

| T 12 | Name: Solution | |

In the given standard orthographic views of the 3D solid object add the missing lines without changing contours.

| T 13 | Name: Solution | |

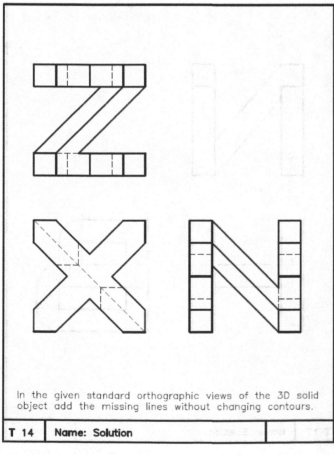

In the given standard orthographic views of the 3D solid object add the missing lines without changing contours.

| T 14 | Name: Solution | |

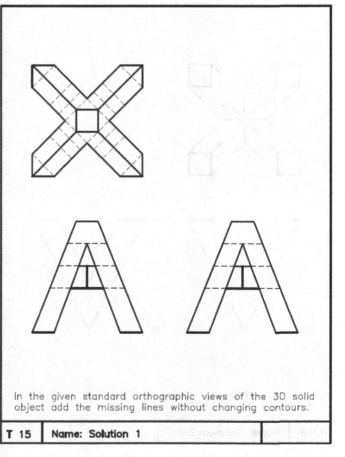

In the given standard orthographic views of the 3D solid object add the missing lines without changing contours.

| T 15 | Name: Solution 1 | |

In the given standard orthographic views of the 3D solid object add the missing lines without changing contours.

| T 15 | Name: Solution 2 | |

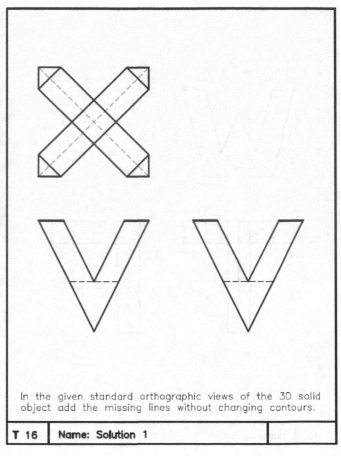

In the given standard orthographic views of the 3D solid object add the missing lines without changing contours.

| T 16 | Name: Solution 1 | |

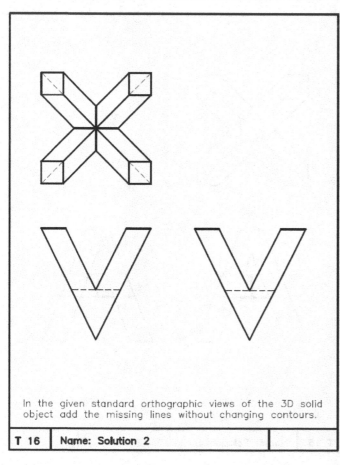

In the given standard orthographic views of the 3D solid object add the missing lines without changing contours.

| T 16 | Name: Solution 2 | |

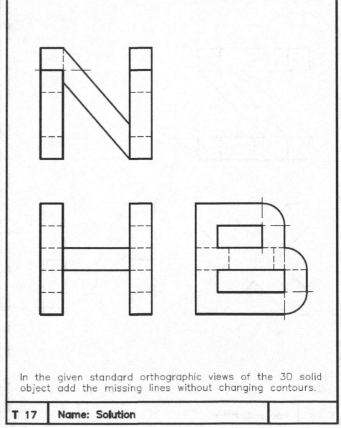

In the given standard orthographic views of the 3D solid object add the missing lines without changing contours.

| T 17 | Name: Solution | |

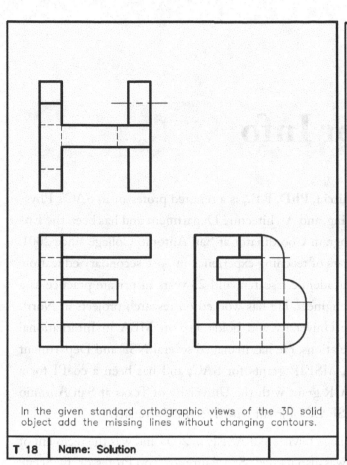

In the given standard orthographic views of the 3D solid object add the missing lines without changing contours.

| T 18 | Name: Solution | |

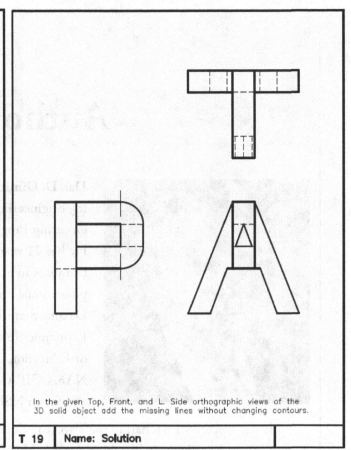

In the given Top, Front, and L. Side orthographic views of the 3D solid object add the missing lines without changing contours.

| T 19 | Name: Solution | |

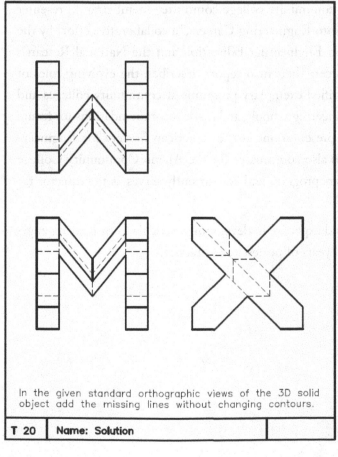

In the given standard orthographic views of the 3D solid object add the missing lines without changing contours.

| T 20 | Name: Solution | |

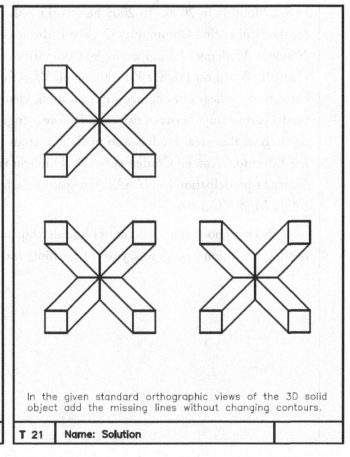

In the given standard orthographic views of the 3D solid object add the missing lines without changing contours.

| T 21 | Name: Solution | |

Author Info

Dan D. Dimitriu, PhD, P.E., is a tenured professor in SAC's Physics, Engineering, and Architecture Department and has been the Engineering Program Coordinator at San Antonio College since 2001. He has 21 years of teaching experience in post-secondary education, five years in academic research, and 23 years in private practice as a professional engineer. He has worked on research projects at North Dakota State University and holds also an MBA in International Economic Relations. He has managed several NSF and Department of Education MSEIP grants for SAC, and has been a co-PI for a NASA CIPAIR grant with the University of Texas at San Antonio and for an NSF CCLI grant with Wright University.

He was elected Vice Chair of the Two Year College Division of ASEE in 2005 and was the recipient of 2006 NISOD Excellence in Teaching Award. He was also named "San Antonio's Top Professor" by Scene in SA Monthly in 2006. In 2005 he was the only community college committee member and presenter for the "Enhancing Community College Pathways into Engineering Careers," a collaborative effort by the National Academy of Engineering's Committee on Engineering Education and the National Research Council Board on Higher Education and Workforce . Their final report described the evolving roles of community colleges in engineering education, identified exemplary programs at community colleges and model partnerships between two- and four-year engineering schools, and made recommendations for future research in this area. He has also made numerous presentations at the American Society for Engineering Educators Annual Conferences. Dr. Dimitriu is also coordinator for the Alamo Community College District's participation in NASA's Aerospace Scholars program and concurrently serves as the director for SAC's MESA Center.

This workbook is a resultant of his leadership and expertise in developing curricula, coordinating engineering educational programs, years of teaching, and years of professional practice.

Printed in the United States
by Baker & Taylor Publisher Services